Was Einstein Right?

WAS EINSTEIN RIGHT?

Putting General Relativity to the Test

CLIFFORD M. WILL

BasicBooks
A Division of HarperCollins*Publishers*

Library of Congress Cataloging-in-Publication Data

Will, Clifford M., 1946–
Was Einstein right?

Bibliography: p. 259.
Includes index.
1. General relativity (Physics) 2. Astrophysics.
1. Title.
QC173.6.W55 1986 530.1'1 85–73877
ISBN 0–465–09088–5 (cloth)
ISBN 0–465–09087–7 (paper)

Printed in the United States of America
Designed by Vincent Torre
88 89 90 91 MPC 9 8 7 6 5 4 3

To Leslie, Betsy, and Rosalie

Contents

	Photographs	*following p. 146*
	Preface	*ix*
1.	The Renaissance of General Relativity	*3*
2.	The Straight Road to Curved Space-Time	*19*
3.	The Gravitational Red Shift of Light and Clocks	*42*
4.	The Departure of Light from the Straight and Narrow	*65*
5.	The Perihelion Shift of Mercury: Triumph or Trouble?	*89*
6.	The Time Delay of Light: Better Late Than Never	*108*
7.	Do the Earth and the Moon Fall the Same?	*135*
8.	The Rise and Fall of the Brans-Dicke Theory	*147*
9.	Is the Gravitational Constant Constant?	*160*
10.	The Binary Pulsar: Gravity Waves Exist!	*181*
11.	The Frontiers of Experimental Relativity	*207*
12.	Astronomy after the Renaissance: Is General Relativity Useful?	*225*
	Appendix	*245*
	Suggestions for Further Reading	*259*
	Index	*263*

Preface

Quasars. Cosmic fireball radiation. Pulsars. Black holes. Gravitational lenses. What do these things have in common?

First, they were all discovered after 1960, during a period of unparalleled advances in the technology of scientific investigation, especially in astronomy.

Second, they have attracted intense popular interest. Just look at the success in recent years of the books (*The First Three Minutes*), movies (*The Black Hole*), and television productions ("Cosmos") that have presented them to the general public, to say nothing of the wristwatches (Pulsar) and television sets (Quasar) that carry some of their names.

Third, their existence makes us ask the question, "Was Einstein right?" Every item in the preceding list involves Einstein's general theory of relativity in a crucial way. Black holes, the remains of dead, collapsed stars, are an important prediction of the theory; a black hole is thought to be responsible for the astronomical X-ray source Cygnus X1, and they are believed by many to power quasars, the incredibly luminous beacons that we can see almost to the edge of the visible universe. The cosmic fireball radiation is most likely the afterglow of the big bang that began the universe, an event whose understanding requires the theory of relativity. The structure of pulsars, believed to be rapidly spinning neutron stars, is strongly influenced by super-strong general relativistic gravitational forces. Finally, the recently discovered gravitational lenses are galaxies that bend and focus passing light by means of the general relativistic warping of space-time around them.

Modern day astronomers and astrophysicists must use general relativity as a tool in their attempts to comprehend these phenomena. If the theory were incorrect, they would be at a loss; an important underpinning of their models would be weakened.

Of course, there is more at stake in the question "Was Einstein right?" than keeping astrophysicists happy (and employed). General relativity is a fundamental theory of the nature of space, time, and gravitation, and has profoundly influenced how we view the universe. But like any theory of nature, it cannot stand on its own. It must face the test of experiment and observation. No matter how profound it may be, no matter how beautiful or elegant it may appear, it must be discarded if it does not agree with observation. Unfortunately, observations of quasars, pulsars, and the like don't in themselves tell us much about general relativity. The reason is that these objects involve such complex physics that we can't easily distinguish the effects of general relativity from the other forces at work. So to find out if Einstein was right, we must look at different kinds of tests.

This book is about those tests. It is about an intensive twenty-year effort, beginning around 1960, to check the predictions of general relativity accurately, and to find new predictions to check. Testing Einstein's theory has always been important, right from the time Einstein published it in 1915, but the string of astronomical discoveries that began with quasars in 1960 gave great urgency to the program. It was not an effort by any individual or team; it was an international effort carried out in laboratories across the United States, Europe, and the Soviet Union. It involved experimental physicists, radio astronomers, spacecraft trackers, theoreticians, and many others. It was not particularly planned or organized, except for periodic conferences that brought the various experts together to report their results. It happened because of three factors: the motivation provided by the kinds of astronomical discoveries that I previously listed; a revolution in the technology of physics and astronomy that provided the tools to do the job; and major theoretical advances in our ability to calculate and interpret the observable consequences of general relativity.

Einstein would not recognize much of what I will describe in this book. The new scientific technologies of the 1960s and 1970s would of course be unknown to him. Hydrogen maser atomic clocks, radar tracking of planets, radio interferometry of quasars—each was discovered or developed well after his death in 1955, and each was used to retest one of the three famous predictions that he made when he first created general relativity. But he would also not recognize the new tests. Two additional and important verifications of his theory were not even discovered as theoretical possibilities until the middle 1960s and were not completed until the 1970s. A third new test, finished just in time for the 1979 centenary year of his birth, confirmed a prediction of the theory about whose reality Einstein himself sometimes harbored doubts.

Would Einstein be pleased with the results? I think he would, since his theory has passed every one of the tests with flying colors.

Would he be impressed? I'm not so sure. It's not that he had no appreciation for experimentation; on the contrary, he coauthored several experimental papers in his early days, and had numerous patents to his credit. He also made it a point to propose observational tests of the general theory of relativity. Yet he appeared to be blasé about the actual outcome of those experiments. In 1930 he wrote, "I do not consider the main significance of the general theory of relativity to be the prediction of some tiny observable effects, but rather the simplicity of its foundation and its consistency." He was apparently completely convinced of the validity of general relativity and believed that experiment would simply confirm what he already knew.

For better or worse, we of the late twentieth century have become more cynical. Over and over we have seen beautiful theories put together, only to be shot to pieces by a conflict with experiment. Just because general relativity has been around for seventy years does not mean that it should not be treated with the same skepticism as are, say, modern theories of the elementary particles. Besides, since gravitation is the oldest known, and in many ways most fundamental force of nature, does it not deserve an empirical foundation second to none?

One subject that this book is not about is the special theory of relativity. While general relativity is a theory of gravitation and of space-time in the large, special relativity deals specifically with physics in the absence of gravitation, and is most important for physics in the small scales of atoms and elementary particles. The reason for its absence here is that special relativity is now a totally accepted and integrated part of modern physics, a basic ingredient in our picture of atomic structure, the atomic nucleus, elementary particles, everything in the world of the microscopic. Experimentally, there is simply no doubt about its validity; it has been checked and rechecked and confirmed time and time again. Nevertheless, a brief description of the main ideas of special relativity and its empirical basis is provided in the appendix.

Although this book is about physics, it is also about people, for it is people who try to understand the predictions of Einstein's theory and who conduct the experiments and observations to test those predictions. Throughout this book I have tried to illustrate the personal side of this endeavor, using my own recollections and the recollections of colleagues.

For their invaluable assistance, I want to thank John Anderson, Francis Everitt, William Fairbank, Russell Hulse, Kenneth Nordtvedt, Irwin Shapiro, and Robert Vessot. I am grateful to Bernard Schutz and Michael Friedlander and to Richard Liebmann-Smith of Basic Books for reading the manuscript and making important and helpful suggestions. Special thanks go to Irwin Shapiro who generously performed a detailed, paragraph-by-paragraph critique of some two-thirds of this book. He apologized for nit-picking, but his comments helped me avoid numerous inaccuracies and errors. For any remaining errors or omissions, I am ultimately responsible.

Was Einstein Right?

1

The Renaissance of General Relativity

SEPTEMBER 14, 1959 was a Monday. It was the beginning of my second week of high school. A diminutive, twelve-year-old ninth grader, I lugged around a briefcase that was half as tall as I was. I was very interested in the academic subjects that I was taking—English, history, mathematics, French, science—but I was especially excited about drafting, the subject most directly related to my professional objectives: I had decided to be an architect, primarily because I had heard that they made a lot of money. I was not especially interested in science, was totally unaware of general relativity, and knew the name "Einstein" only as the generic term for genius. "That person is a real brain, another Einstein," was the sort of thing we used to say. I was completely oblivious to the gathering winds of change in the science of general relativity that would pick me up and carry me along a decade later.

On September 14, there was no way that I could have been aware of what was happening in the 26-million-mile void somewhere between the Earth and Venus. Twelve days earlier, in the closing moments of my summer vacation, Venus had passed through that point in its orbit at which it is the closest to Earth, called "inferior conjunction," and scientists of the Massachusetts Institute of Technology's Lincoln Laboratory were trying to take advantage of this proximity to perform an important experiment. Their goal was to send a radar signal to Venus, allow it to reflect off the surface of the planet, and attempt to detect the radar echo that would return roughly four and a half minutes later. For a four-week period in late August and September they sent coded radio waves toward Venus from MIT's Millstone Hill radar antenna, located 20 miles northwest of Boston, and waited anxiously for echos. Unfortunately, the returning signals were extremely weak, and to their disappointment, the scientists were unable to detect any echoes above the inevitable noise that is present in any radio receiver. As Venus receded from the Earth, the scientists stopped sending signals and prepared to wait for the next inferior conjunction of Venus, due on April 9, 1961.

The April 1961 series of radar observations were much more successful; echoes from Venus were detected without difficulty, and the round-trip travel time for each signal was measured with good accuracy. When combined with the speed of light, these measurements gave the distance between the Earth and Venus at that time to about a 100-kilometer accuracy. Similar success was reported at the Jet Propulsion Laboratory of the California Institute of Technology. Echos were also detected at radar installations in England and the Soviet Union. An important discovery was made during these observations: The accepted value of the "astronomical unit," the quantity that fixes the size of the Earth's orbit around the Sun, about 150 million kilometers, was in error, too small by about six-hundredths of a percent, or 93,000 kilometers. Because the astronomical unit is the basic unit of length for determining all planetary orbits, such a correction would change the calculated orbits correspondingly. While analyzing the 1961 data, the Lincoln Laboratory scientists realized that this correction in the astronomical unit might have been responsible for the failure of the earlier observations to detect echoes.* The 1959 observations had been made using the old value of the astronomical unit to determine the expected return time of the echo, and as a consequence scientists had been looking for the echo in the wrong place. Sure enough, when the data from the September 14, 1959 observations were reanalyzed using the corrected astronomical unit, an echo was found.

Although the detection of an echo in the September 14 data was done as an afterthought, the observations of that day served as the event that opened a remarkable academic year—September 1959 to September 1960—a year with great portents for Einstein's general theory of relativity.

What else happened that year?

Physical Review Letters is a scientific journal that publishes physics papers that are deemed to be of such importance that rapid publication is called for in order to disseminate the new results as quickly as possible to the scientific community. On March 9, 1960, the editorial office of the journal received a paper by Robert V. Pound and Glen A. Rebka, Jr., of Harvard University, entitled "Apparent Weight of Photons." The paper described the first successful laboratory measurement of the change in the frequency or wavelength of light as it falls in a gravitational field. The phenomenon is called the gravitational red shift of light, and was the first of the three famous predictions Einstein made using general relativity. The Pound-Rebka paper was accepted and published in the April 1 issue.

A few months later, in the June 1960 issue of another physics journal known as the *Annals of Physics,* there appeared a paper by the English mathematical physicist Roger Penrose with the esoteric title "A Spinor Approach to General Relativity." Although the paper was highly mathematical, it outlined a very elegant and streamlined calculational technique for solving

* When attempting to find a weak signal buried in noise, it helps to know where to look (the classic "needle in a haystack" problem). In the case of a weak radar echo, we want to know approximately when the echo is expected, so that we can zero in on that part of the data.

certain problems in general relativity. The theory had long had the reputation of being extremely difficult to deal with mathematically, yet this new method made some computations almost easy.

Later that summer, while I was playing schoolyard baseball, going on hikes in the forests near my home, and pondering whether I should forget architecture and become a geneticist, Carl H. Brans was beginning to put the finishing touches on his Ph.D. thesis. Brans was a graduate student at Princeton University, working under the eminent experimental physicist Robert H. Dicke. Yet his thesis was devoted to theory. It was entitled "Mach's Principle and a Varying Gravitational Constant." A portion of the thesis presented the equations for a new theory of gravity, an alternative to Einstein's general relativity. He referred to the new theory as a "scalar-tensor" theory of gravity, but in time it came to be known as the Brans-Dicke theory.

As fall approached, Brans continued working on his thesis, I began the tenth grade, now absolutely convinced that genetics was my calling, and astronomers Thomas Matthews and Allan Sandage prepared to use the 200-inch telescope at Mt. Palomar in California to make some observations of a radio source denoted 3C48 (the forty-eighth entry in the third "Cambridge catalogue" of radio sources). They were interested in what kind of visible light this source might be emitting, so on the night of September 26, 1960, they took a photographic plate of the area of sky around 3C48. Conventional wisdom at the time told them that they would find a cluster of galaxies at the location of the radio source, but this was nothing like what they saw. Instead, as far as anyone could tell by looking at the photographic plate, the object was a star. Yet it was like no other star seen up to then, for subsequent observations during October and November of that year and periodically throughout 1961 showed that its spectrum of colors was highly unusual, and that its brightness or luminosity varied widely and rapidly, sometimes over periods as brief as fifteen minutes. This was a new addition to the astronomical family, and it needed a special name. It was a radio source, yet it looked "stellar" or starlike (ordinary stars are not strong radio sources); on the other hand, because of its spectrum and variability it was not quite a star, it was only "quasi" stellar. Hence the name quasi-stellar radio source or "quasar" was soon applied to this object and to others like it.

This remarkable discovery concluded the academic year 1959–60, just over a year after the radar observations of Venus at inferior conjunction. It was a remarkable year for general relativity, because it contained all the signs that a renaissance was about to begin. The five disparate and seemingly unrelated events that I have described, in fields ranging from experimental physics to abstract mathematics to astronomy, set the stage for an era in which general relativity would become an active and exciting branch of physics, after almost a half century in the backwaters. It was to be a period in which general relativity not only would become an important working tool of astrophysicists in their attempts to unravel the mysteries of astronomical phenomena, but also would have its validity called into question as never before. Yet it was also to be a time in which experimental tools would become available to test and confirm the theory in unheard-of ways and to incredible levels of precision.

The discovery of quasars thrust general relativity immediately to the forefront of astronomy. The reason was an energy crisis of truly cosmic proportions. Within a few years after the discovery of the quasar associated with 3C48, it was found that it and other quasars like it were among the most distant objects in the universe. What the astronomers thought were unusual spectra were actually rather ordinary spectra in which all features were shifted uniformly to the red end of the spectrum. This meant that the quasars must be moving away from us at high speed, 30 percent of the speed of light in the case of 3C48. The shift in wavelength to the red is a consequence of the Doppler shift, the same effect that, in the case of sound, causes a railroad crossing bell to have a lower pitch when we are moving away from it. In 1929, the astronomer Edwin Hubble had discovered that the systematic "red shift" of the spectra of distant galaxies, first observed fifteen years earlier by Vesto M. Slipher, had a pattern associated with it, so that the farther the object, the larger its recession velocity (in almost direct proportion), and the larger the red shift of its spectral features. Since that time, it was accepted that the universe is expanding uniformly in all directions. For 3C48, for instance, the recession velocity of one-third of the speed of light corresponded to a distance of about 6 billion light years. The light that we see today from 3C48 left it when the universe was two-thirds as old as it is now. Because the quasars were so distant, one would have expected them to be faint, yet they were very bright sources, both in visible light and in radio waves. Therefore their intrinsic brightness or luminosity must be enormous. For 3C48, the numbers translated into 100 times the brightness of our own galaxy.

This was the energy crisis: What could possibly be the source of such power? On cosmic scales, the strongest force known is gravity, so it was suggested that the energy of super-strong gravitational fields could provide the answer. Furthermore, the source of this power had to be very compact, for the simple reason that for the source to vary in brightness coherently over a period of, say, one hour, it couldn't be much larger than the distance light can travel in one hour, in order for one side of the source to know what the other side is doing and thus to behave in unison.

Thus, one solution to the quasar energy crisis involved strong gravitational fields, meaning perhaps a huge concentration of mass, maybe millions of times the mass of the Sun, confined to a region of space smaller than a light hour or about the diameter of the orbit of Jupiter. This represented a new collapsed state of matter that could only be described by the general theory of relativity.

The discovery of quasars created a new field of physics. In June 1963, invitations were sent out to astronomers, physicists, and mathematicians around the world to attend a conference on this new discipline, to be called Relativistic Astrophysics. The First Texas Symposium on Relativistic Astrophysics was held in Dallas, on December 16–18, 1963, with some three hundred scientists in attendance. The atmosphere was electric, partly because of the all too recent shock of the assassination of President Kennedy in that city just three and a half weeks earlier, and partly because many of the participants sensed the enormous potential of this new field for confronting the most fundamental questions of nature. The fact that astronomers, physicists, and mathematicians were being brought together to work in concert on these kinds of questions was also tremendously exciting, although at first it had its amusing side. Several participants at that first Texas symposium tell of the general relativity theorist interrupting a lecture by an astronomer to ask what he meant by the "magnitude" of a star (magnitude is the astronomer's measure of the brightness of a star, an elementary concept taught in every freshman astronomy class), or of the astronomer asking the general relativist what the "Riemann tensor" was (the Riemann tensor is a measure of the curvature of space-time, to the relativist an equally elementary concept). But soon the practitioners of this new interdisciplinary field learned how to communicate with each other, so that by later Texas symposia, it was not uncommon to find relativistic astrophysicists who were as knowledgable about the intricacies of curved space-time as they were about the structure and evolution of stars or about the capabilities and limitations of radio telescopes.

Yet all this interdisciplinary effort raised the question, "Is general relativity the correct foundation for relativistic astrophysics, or should some other theory be used?" Within a few years, the discovery of the cosmic fireball radiation (1965), of pulsars (1967), and of an X-ray source that might contain a black hole (1971), made it even more urgent to find an answer.

Roger Penrose's paper in the 1960 *Annals of Physics* served as the announcement that a new school of general relativists was on the scene. Prior to this time, general relativists had the reputation of residing in intellectual ivory towers, confining themselves to abstruse calculations of formidable complexity, with results that were impossible to understand. But the new relativists, represented by people like Penrose, did two things that were important for the future of the subject. First, they brought with them new mathematical techniques that allowed them to simplify and streamline many calculations, making some of them almost trivial. This also helped to wash away some of the mystery that had plagued the subject and made the teaching of general relativity to new generations of physicists a simpler and more successful pastime. Second, they grasped the importance of focusing on the physically observable or detectable consequences of the theory, rather than becoming bogged down in mathematical subtleties. These developments helped bring general relativity back into the fold of mainline physics. This was a far cry from the days when it was said that only a dozen people in the world understood the general theory of relativity.

Actually, the origin of the notion of the theory's inaccessibility is unclear. The headline of an article on relativity in the November 9, 1919 issue of the *New York Times* stated, "A book for 12 wise men/ No more in all the world could comprehend it, said Einstein when his daring publishers accepted it." Einstein himself may have used some such phrase as early as 1916 in reference to a popular book on relativity that he had written. However, my favorite story about this myth has it originating with the British astronomer Sir Arthur Stanley Eddington, or so I was once told by the relativistic astrophysicist Subrahmanyan Chandrasekhar, who worked with Eddington in the 1930s. Soon after the publication of the final form of general relativity in 1916, Eddington was one of the first to appreciate its importance, and set out to master it (in this sense he was the first relativistic astrophysicist). He also participated in the 1919 expedition to photograph the Sun during a total solar eclipse in order to see if the light from stars was bent upon passing by the Sun in the amount predicted by the theory. At the close of the meeting at which the successful results of the eclipse expedition were reported, a colleague said, "Professor Eddington, you must be one of three persons in the world who understands general relativity!," to which Eddington demurred. The colleague persisted, saying, "Don't be modest, Eddington." Eddington replied, "On the contrary, I am trying to think who the third person is."

Perhaps only three people understood it, but millions were certainly fascinated by it and wanted to read about it and about Einstein. In the popular press, the scientific revolution engendered by general relativity was placed on a par with the insights of Copernicus, Kepler, and Newton. Editorial after editorial marveled at what was called one of the greatest achievements in the history of human thought, but at the same time complained about the difficulty of understanding it. Einstein himself wrote a long article for the London *Times* in late 1919, attempting to explain the theory to a general audience. His picture graced the cover of the December 14, 1919 issue of the German newsmagazine *Berliner Illustrirte*, over the caption, "A new great figure in world history." The Einstein legend was born, and has not subsided to this day.

It is ironic that while the legend of Einstein and his theory was growing, the actual science of general relativity was becoming stagnant and sterile. By the middle 1920s Einstein had turned most of his attention to the futile quest for a unified field theory that would combine gravitation and electromagnetism, and many other researchers followed suit. With only a few exceptions, most work in general relativity during the next thirty-five years was devoted to abstract mathematical questions and issues of principle, and was carried out by a small band of practitioners. The relativist Peter G. Bergmann once said of this period, "You only had to know what your six best friends were doing, and you would know what was happening in general relativity." Until 1955, there was not a single international conference devoted exclusively to general relativity.

By 1960, the attitude seemed to be that while general relativity was undoubtedly of great importance as a fundamental theory of nature, its observational contacts were severely limited, understanding it was difficult, and doing calculations with it was nearly impossible. A classic illustration of this feeling is the story of a recent graduate of the California Institute of Technology who asked advice of his professors as to what he should specialize in during his upcoming graduate work at Princeton. He was told by a famous Caltech astronomer that under no circumstances should he do general relativity, because it "had so little connection with the rest of physics and astronomy." This was 1962. Fortunately for the field (and for me), the student, whose name was Kip Thorne, paid no heed. Like Penrose, he became one of the new breed of relativists, helping to shatter the old myths about the subject. Within four years he was back at Caltech as a professor, embarked on setting up a major research center on relativistic astrophysics.

Younger theorists like Penrose and Thorne, like Stephen Hawking, James Hartle, Igor Novikov, and James Bardeen, together with some young-at-heart converts to the new relativity, like John Wheeler, Subrahmanyan Chandrasekhar, Alfred Schild, and Yaakov Zel'dovich began to take over and dominate the field. Armed with the new calculational tools and the desire to apply the theory to the real world, they began to make the important theoretical discoveries in general relativity that had been missed for four decades. Spurred on by the new astronomical findings, they started to work out the theory of relativistic neutron stars, the properties of black holes, the nature of gravity waves, and the evolution of the universe from the big bang to the present. They also wrote the new general relativity textbooks and popular articles that would spread the exciting word to new generations of relativists, as well as to the general public.

All this new research was based on the assumption that general relativity was correct. This assumption took on added importance because of another of the events of 1959–60: the invention of the Brans-Dicke theory of gravity.

The Brans-Dicke theory provided a viable and attractive alternative to general relativity. Its very existence and agreement with all observational evidence to that point demonstrated that general relativity was not a unique theory of gravity. Some even preferred the new theory on aesthetic and theoretical grounds. To be sure, it was not the first alternative to general relativity ever invented. There had been many others, some devised even before general relativity as early attempts to forge a unification of gravity with Einstein's special theory of relativity, others devised later as responses to the supposed complexity and mystery surrounding general relativity. But none of these was taken quite as seriously as was the Brans-Dicke theory. This was because the theory retained many of the features of general relativity that most theorists viewed as essential, but added some new features that general relativity lacked. In most cases, its observable predictions were different, but not radically different from those of Einstein's theory. The theory also began to be taken more seriously in the middle 1960s when results from a certain experiment appeared to go against general relativity.

Experiment. That was the key. Without experiment, physics is sterile, physical theory merely idle speculation. Any theory must stand or fall ultimately based on its agreement or disagreement with experiment. Unfortunately, observations of quasars or pulsars do not generally give us good experimental tests of gravitation. These astronomical phenomena are usually so complex that it is difficult if not impossible to disentangle the gravitational effects from all the other physical processes in order to see if the gravitational predictions hold up. Instead, we need experiments that are somewhat closer to home, where life is simpler and cleaner, where we can hope for a measure of control over the experimental conditions. Although Einstein was not particularly motivated by experimental results when he set out to develop general relativity, he was well aware of this need for experimentation, and proposed three tests of general relativity, known as the deflection of light by the Sun, the perihelion advance of Mercury, and the gravitational red shift of light. The first two were successful; the third was inconclusive. However, even the two successes were only partial. The effects predicted by general relativity were observed, but the accuracies were low. By 1960, this was simply not good enough. The Brans-Dicke theory also predicted these effects, but the sizes were slightly different. Therefore, the mere detection of a relativistic effect was not enough—what was now required was a new set of high-precision experiments, with the ability to measure these effects (which as we will see are always very tiny) to accuracies of 10 percent, 1 percent, or fractions of a percent, in order to distinguish one competing theory from another. The Brans-Dicke theory forced general relativity to confront experiment as never before.

But to do high-precision experiments, we need tools, and the remaining two major events of 1959–60 signaled that the needed tools were on the way. The 1960 Pound-Rebka experiment, besides giving the first verification of Einstein's third prediction, the gravitational red shift of light, demonstrated the powerful use in precision experiments of the new technologies that would explode during the 1960s and 1970s. These new technologies, based on quantum mechanics, semiconductors, masers and lasers, superconductivity, computers, and so on, would provide the tools to test Einstein's theory to levels of accuracy that would have been unimaginable just ten years earlier. The remaining event, the one that started this remarkable year for general relativity, provided us with the laboratory for many of the experiments. Sending radar signals to Venus (and ultimately recording their echos) opened up the solar system, not only as a place for planetary exploration and for the search for extraterrestrial life, but also as an arena for testing general relativity. The rapid development during the early 1960s of the interplanetary space program made techniques such as radar ranging to planets and satellites a vital new method for probing relativistic effects. The names Mariner, Apollo, Viking, usually associated with the excitement of close-up pictures of Mars and of men walking on the Moon, would also become part of the language of experimental relativity. For the next decade and a half, until the summer of 1974, the solar system would be the central arena for testing whether Einstein was right.

By the time of the 1979 centenary of the birth of Einstein, the relativistic renaissance begun in 1960 was in full swing. The outpouring of books commemorating that great event attested to the vigor and excitement of research in relativistic astrophysics. Black holes were an accepted and understood part of the theory, and the observational evidence for their existence was compelling. A model for the basic structure and evolution of the universe was in good shape, and cosmologists were beginning to explore what might have happened in the first trillionth of a second after the big bang. The nature of quasars ironically remained a mystery after twenty years, while a pretty good theory of pulsars was in hand. As a preview of the centenary year, in December 1978, the Ninth Texas Symposium on Relativistic Astrophysics, held for the first time outside the United States, attracted over eight hundred participants to Munich, Germany. All these recent developments, and numerous others were discussed by the various lecturers.

But whereas the first Texas symposium in Dallas in 1963 had not included a single lecture about the experimental verification of general relativity, the ninth featured two. One lecture, by Joseph H. Taylor of the University of Massachusetts, described the latest result from a remarkable new testing ground for general relativity that had been discovered in 1974. This new arena for experimental relativity was a pulsar in orbit about a companion star, colloquially called the binary pulsar. Taylor reported how observations of the orbit of the pulsar since 1974 had led to the first confirmation of one of the most important predictions of Einstein's theory: the existence of gravitational waves. The other lecture, by me as it happens, told how the effort to verify general relativity, spurred by the events of 1959–60, had become an active and challenging field, with many new experimental and theoretical possibilities. These included new versions of the original tests proposed by Einstein, such as the deflection of light and the gravitational red shift, with accuracies that were unthinkable before 1960. But they also included brand new tests of Einstein's theory that had been discovered theoretically after 1960. The bottom line was that general relativity had passed every test with flying colors, and that many other theories, including the Brans-Dicke theory were in decline. Yet there was more to be done, further experimental possibilities at the frontiers of observational relativity that needed to be realized in order to continue to strengthen the empirical foundations of Einstein's theory.

What had happened to convert me from oblivious teenage architect/geneticist to exponent of experimental relativity?

The answer is that I was swept up in the relativistic renaissance that was already full blown in the spring of 1969 in the research group that Kip Thorne had built up at Caltech. A list of the subjects that he and his many graduate students were working on at that time reads like a "what's hot" list for relativistic astrophysics: different kinds of cosmological models within general relativity, the collapse of stars to form black holes, the vibrations of neutron stars and the emission of gravitational radiation, the structure of relativistic rotating neutron stars, and super-dense clusters of stars as possible models for quasars. Along with John Wheeler and Charles Misner, Thorne was also hard at work on the first draft of a textbook on general relativity that would emphasize these modern developments.

But Thorne himself was sufficiently concerned about whether Einstein was right that he decided to assign his latest incoming young graduate student to look into it. From that point on I was hooked. Although I had long before committed myself to physics, I had entered Caltech undecided on a specialty. But I found myself drawn to "Kip's" group by the aura of excitement that emanated from it, by the feeling that they were in the thick of a great adventure in science, and by the sense that, in addition to anything else, it was just plain fun. I believe that what made this new relativity fun compared to the "old" relativity was the mixture of abstract theoretical computation with the ability to compare the results with real observations. You could play with lofty space-time concepts and at the same time "get your hands dirty" with actual data. For instance, my notes from this period include highly mathematical calculations of potentially observable effects on the motions of bodies predicted by general relativity as well as by other theories of gravity. But from the same period, I also have notes from discussions I had with scientists from the nearby Jet Propulsion Laboratory on what radar tracking accuracy NASA would provide on a proposed mission to Mars called Viking, and whether it would be possible to test a relativistic effect known as the "time delay" with enough precision to distinguish between general relativity and the Brans-Dicke theory.

However, it is just possible that the process of conversion for me began much earlier. By 1961, I had pretty much dumped genetics as a career choice, and was thinking more along the lines of physics. In December 1961, I had my first encounter with experimental gravitation, and with a name that will appear in many places in this book: Robert H. Dicke. Just as Dicke had a tremendous impact on the field of testing general relativity, he also had an effect on me, although at the time it was surely subliminal. I had received a gift subscription to *Scientific American* for my fifteenth birthday, beginning with the December 1961 issue. Naturally, like any new subscriber to such a magazine, I read that first issue from cover to cover, with great interest, and almost no understanding. It was exciting to read about "The East Pacific Rise," the "Three Dimensional Structure of a Protein," and the "Prehistoric Swiss Lake Dwellers," but in twenty-five years, until I looked up that issue while writing this chapter, I had totally forgotten about all those wonderful articles. All but one.

It was an article by Dicke entitled "The Eötvös Experiment." I don't know why this article has always stuck with me. I certainly didn't understand it, any more than I understood the others. Was it something having to do with the Hungarian baron, Roland von Eötvös, who first performed this experiment at the turn of the century? Was it the name Princeton, at whose university Dicke was attempting to improve upon the experiment, and the location of the Institute for Advanced Study where Einstein spent the second half of his scientific career? Was it because Dicke claimed that somehow this mundane-looking apparatus with wires and mirrors and vacuum tubes and weights could tell us that space-time was curved? Was this article trying to tell me something truly fundamental?

Now, rereading the article twenty-five years later, I still feel a bit of the wonder and mystery that I experienced then. I now understand the article completely, of course. Dicke's experiment was not completed until a few years after that article appeared, and its results, like those of Eötvös over sixty years earlier, do tell us something fundamental, more fundamental than general relativity itself. They tell us that space-time is curved, whether we believe Einstein's version or the Brans-Dicke version or some other version. So before we embark on our journey to answer the question, "Was Einstein Right?," let us begin with a straight section of road to curved space-time.

2

The Straight Road to Curved Space-Time

WHEN space-shuttle astronaut Sally K. Ride was a graduate student at Stanford University in the middle 1970s, she somehow missed the opportunity to take my course on Einstein's theory. By the time I taught it, she had completed most of her course work, and was deeply involved in the research in X-ray astronomy that would form her Ph.D. thesis. Yet ultimately she had a closer and more personal experience of the concept underlying all of gravitation than I ever will. She experienced weightlessness, or the apparent disappearance of gravity, that occurs inside any orbiting or freely falling craft.

After almost a quarter of a century of seeing orbiting astronauts on the nightly television news, we take the idea of weightlessness rather for granted, and now think mainly of its physiological effects on the astronauts, or of its possible industrial or pharmaceutical applications. But to Einstein, the idea was a profound one, for to him it meant that space-time must be curved.

The proposal that space-time must be curved was a product

of Einstein's genius for taking a simple experimental observation, combining it with an idealized imaginary experiment (called a *gedanken* or "thought" experiment) that incorporates the essence of the original experiment, and pushing the result to its logical limit. In this case the observational result was the commonplace one that bodies fall with the same acceleration. This brings to mind the image of Galileo Galilei dropping objects from the top of the Leaning Tower of Pisa.

Einstein took this simple observation, and imagined what it would imply for an observer inside an enclosed, freely falling laboratory. Of course, in 1907, when Einstein first began to ponder this question, it had to be a pure thought experiment, for the dawn of the space age was still fifty years to the future, and the flight of Sally Ride another twenty-six years after that. Nevertheless, the weightlessness, or vanishing of gravity, that such an observer would experience seemed so significant to Einstein that he elevated it to the status of a principle, which he called the principle of equivalence. "Equivalence" came from the idea that life in a freely falling laboratory was equivalent to life with no gravity. It also came from the converse idea that a laboratory in distant empty space that was being accelerated by rockets was equivalent to one at rest in a gravitational field. From this equivalence, Einstein concluded that space-time must be curved. Before we attempt to understand this remarkable conclusion about space-time, let us cover the more familiar ground of space. Most of us are familiar with the Euclidean picture of space. At one time or another, we have plowed (or have been forced to plow) through Euclid's books and theorems on plane geometry and have learned, among other things, that parallel straight lines never intersect, that the interior angles of a triangle sum to 180°, and so on. We are comfortable with these ideas because they match our everyday experience with sheets of paper and table tops. The kind of space described by the postulates of Euclid is called a flat space.

Actually, some kinds of curved spaces can be understood very easily. One example is the surface of a sphere (see figure 2.1). It violates Euclid's postulates, because on the surface of a

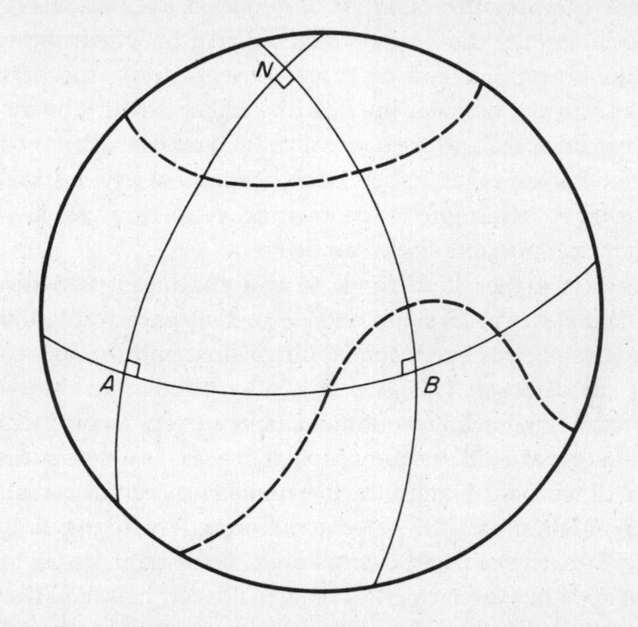

Figure 2.1 Curves on a sphere. Solid curves are examples of great circles or geodesics, such as the equator or lines of longitude, in fact any curve formed by the intersection of a plane through the center of the sphere with its surface. Dashed curves are not geodesics. The triangle *NAB* has a total interior angle of 270°.

sphere, the "straightest" lines are the "great circles." Examples of great circles are the lines of longitude and the equator. In any curved space, such lines are called geodesics. But notice that two lines of longitude, which are parallel at the equator, actually intersect at the poles, a clear violation of Euclid's postulates. Similarly, we can form a triangle whose interior angles add up to more than 180°. For instance, consider a triangle made up of the 0° line of longitude, the 90° line of longitude, and the equator. The total interior angle is not 180°, but three right angles, or 270°. Another example of a curved surface is the surface of a saddle, which appears concave in one direction (along the spine of the horse) but convex in the other direction.

These examples are easy to understand because they are spaces of smaller dimension than the one in which we live, two dimensions instead of three. We can look upon these spaces from the outside; they can be imbedded in our three-dimensional space and thus visualized. Of course, even though it is not necessary to imbed such a space in one of higher dimension to determine its geometrical properties, we always feel more comfortable if we can do so.

This fact makes it difficult to visualize a curved three-dimensional space. To stand outside such a space and visualize it would require a fourth spatial dimension, and that is clearly not at our disposal. This in part is why it took two thousand years following Euclid for mathematicians even to consider the question of curved three-dimensional spaces, or even spaces of higher dimension. Finally, in the nineteenth century, mathematicians such as Carl Friedrich Gauss, Wolfgang Bolyai, Nikolai Lobachevski, and Georg Friedrich Riemann began to be able to spell out the properties of such spaces, mathematically, if not pictorially.

One step harder to visualize, of course, is a curved four-dimensional space-time. But why should we worry about space-*time* as opposed to ordinary space? Einstein's special theory of relativity provides the answer. Special relativity replaced the Newtonian concepts of space and of a separate absolute time with a single geometrical framework of space-time, in which the spatial degrees of freedom and the time are treated on a more equal footing, and can be interrelated. For example, the rate of flow of time depends on the state of motion of the observer: a clock in a moving laboratory appears to tick more slowly than a set of identical clocks distributed throughout a reference laboratory. As another example, two events or occurrences at two different locations can be seen to be simultaneous by one observer, but will be seen to be not simultaneous by a moving observer. It was Hermann Minkowski (1864–1909) who developed the idea that a "space-time continuum" was the underlying geometry behind the time and space relationships proposed by special relativity. (A more detailed description of the main ideas and consequences of special relativity is contained in the appendix.)

As difficult as it may be to deal with the four-dimensional space-time of special relativity, at least it is in a way Euclidean. For a given observer, the ordinary three-dimensional spatial part of his world has the properties of a normal flat space.

Yet this is precisely what Einstein proposed: that space-time is curved, not flat, and that this curvature is produced by the gravitational effects of matter. Even more than this, he suggested that, in fact, curvature of space-time is in some sense identical or equivalent to gravitation. When you think about it, this is a truly amazing leap of imagination.

At one time it was thought that gravity was the result of an "action at a distance," and that bodies attracted one another by means of such an action. This idea was Newton's, and forms the underlying basis of Newtonian gravitation. According to this idea, given any two bodies, there was an "action" between them that caused them to attract each other by a force that was proportional to each of their masses and inversely proportional to the square of the distance between them. But by the middle of the nineteenth century, this idea had been supplanted by the concept of the "field." In the field picture, a gravitating body produces around itself a field of force, which exists whether or not a second body is present to feel the force field and be attracted. The field of force is related to a gravitational potential whose variation in space determines the force. Any body possesses a gravitational potential. This "field" or "potential" concept came from the new understanding of electromagnetism that was being developed at this time. For instance, the existence of the magnetic field of a magnet could be demonstrated by sprinkling iron filings on a sheet of paper covering the magnet. The lines of force emanating from the poles could be clearly seen in the pattern of the iron filings. The application of fields to electricity and magnetism carried over quite naturally to gravitation.

What Einstein proposed was a third alternative. In Einstein's view, a gravitating body actually distorts the very fabric of space and time around it. A body that enters the vicinity of the first body merely responds to the distortions of space-time that it encounters.

To see how Einstein could make the leap from the equality of acceleration of bodies and his principle of equivalence to the idea of curved space-time, let us return to the example of a two-dimensional curved space, the surface of a sphere. Imagine a two-dimensional world, much like that of the nineteenth-century book *Flatland* by E. A. Abbott, but here confined to the surface of a sphere. The two-dimensional inhabitants of this "Sphereland" set out to learn something about its geometrical properties by constructing a set of very straight platinum rulers, laying three of them out to form a triangle, and measuring the sum of the interior angles. To their surprise, they discover that the sum is 195°, not the 180° they expected from their knowledge of Euclidean plane geometry. Their initial postulate (remember, they cannot step outside their world to look at it) is that there is a field of force that acts on platinum, causing it to bend in such a way as to make the interior angles exceed 180°. To test this postulate further, they try the same experiment using rulers made of aluminum, with exactly the same result. After trying numerous different substances in their rulers, they conclude that whatever is going on is universal—it affects all rulers equally. Next, the Spherelanders try shorter rulers, and construct a smaller triangle. This time the sum is 187°. Smaller triangles yield a smaller sum of angles. As the triangles get smaller and smaller, the sum of the angles gets closer and closer to the normal value of 180°.

What are the Spherelanders to conclude from this series of experiments? Are there forces in their world that cause rulers to bend in just such a way as to make triangles behave as they did? The fact that all rulers behaved in exactly the same way suggests that the phenomenon has less to do with the rulers themselves than it has to do with the underlying nature of Sphereland. Perhaps, says one imaginative Spherelander, our world is actually curved, not flat; that would explain the triangle experiments, especially their universality. Furthermore, when we confine our attention to progressively smaller regions, the effect of the curvature becomes smaller and smaller, and so the world appears more and more equivalent to a flat world, in the Euclidean sense. Thus, Sphereland is a space that is curved on the large scale but appears approximately flat on the small scale (see figure 2.2). This is, of course, an observation with which we are familiar on the surface of the Earth.

Einstein used a similar line of reasoning to go from the principle of equivalence to curved space-time. Bodies of platinum and aluminum fall with the same acceleration, so perhaps the gravitational force that acts upon them has less to do with the bodies themselves than with the underlying space-time. Just as Sphereland was curved, perhaps here space-time is curved, and the trajectories of falling bodies merely reflect the curves and bends in space-time. Furthermore, we can place ourselves in a freely falling laboratory and discover that we float freely; we seem to feel no gravitational forces. If the laboratory is sufficiently small, then any object that we bring with us into it will float freely with us, as if gravity were absent. This is only an approximation, of course, because we know, for instance, that in a laboratory in free fall above the Earth, the force of gravity at the top of the laboratory is a little weaker than that at the bottom, so there will still be a small residual effect of gravity. These effects are called tidal forces, and are responsible for the tides of the oceans on Earth, for instance. Nevertheless, the smaller we make the laboratory, the smaller these residual tidal forces become, and the closer we get to a completely gravity-free situation. Therefore, in a small, freely falling laboratory, bodies move on straight lines as if gravity were absent, or as if our laboratory were an inertial frame. This is like saying that the space-time inside the laboratory is, at least approximately, the flat space-time of special relativity, just as space in Sphereland

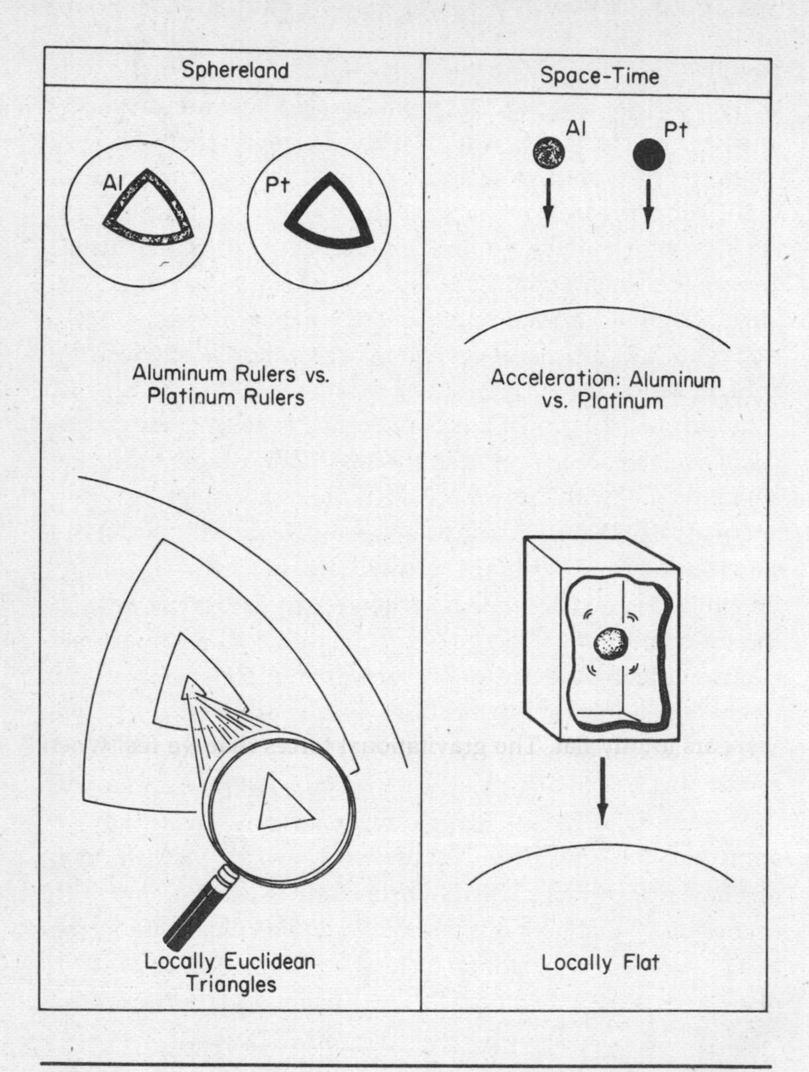

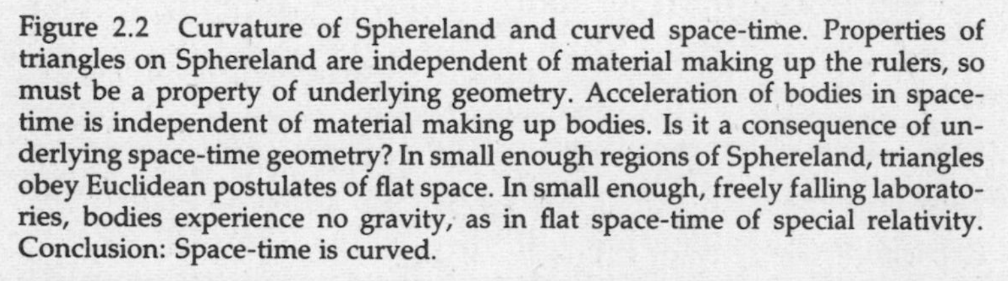

Figure 2.2 Curvature of Sphereland and curved space-time. Properties of triangles on Sphereland are independent of material making up the rulers, so must be a property of underlying geometry. Acceleration of bodies in space-time is independent of material making up bodies. Is it a consequence of underlying space-time geometry? In small enough regions of Sphereland, triangles obey Euclidean postulates of flat space. In small enough, freely falling laboratories, bodies experience no gravity, as in flat space-time of special relativity. Conclusion: Space-time is curved.

on the scale of the smallest triangles was approximately the flat space of Euclid.

Furthermore, Einstein argued, this should apply not only to the mechanical motions of bodies but also to all the laws of physics, including electromagnetism and atomic structure, and to any nongravitational laws that might be discovered in the future (such as quantum mechanics and the laws governing elementary particles). In other words, all the mathematical equations governing such nongravitational laws of physics, when applied to a freely falling frame, were to be written in the form they normally had in the flat space-time of special relativity.

And how do free bodies move in curved space-time? Just as the local straight lines in Sphereland correspond to great circles of the sphere, or geodesics, so in space-time the trajectories of freely falling bodies correspond to geodesics, the "straightest" possible lines.

This was Einstein's leap. The principle of equivalence means that space-time is curved, but that to an observer in free fall, it appears locally flat. The gravitational forces that we feel when we are not in a freely falling laboratory are nothing but a consequence of space-time curvature.

This is not a rigorous theorem, of course. It is not the only possible interpretation of the principle of equivalence. However, it is elegant, it is simple (conceptually, if not mathematically), and it is compelling. Most great theoretical advances are guided by these kinds of criteria. That is not enough to make these advances correct, however. The final arbiter is experiment. In this book we shall encounter many experiments designed to test the consequences of curved space-time and of the specific theory that Einstein constructed to give precise mathematical content to the postulate of curved space-time.

Because the principle of equivalence plays such a crucial role, we need to review its history and its experimental foundations.

Despite the strong influence of the Aristotelian version of mechanics, which held that heavier bodies fall more quickly than light bodies, there were opponents to this view, even in antiquity. For example, Ioannes Philiponos (fifth or sixth century A.D.) recorded, presumably from experience, ". . . if you let fall from the same height two weights of which one is many times as heavy as the other you will see that the . . . difference in time is a very small one." Others who recognized the equality of fall rates included Giambattista Benedetti, who proposed equality in 1553, Simon Stevin, who tested it experimentally in 1586, and Galileo Galilei (1564–1642), whose supposed experiment at the Leaning Tower of Pisa has become part of popular scientific folklore. Galileo took his post as professor at the University of Pisa in 1589, three years after the publication of Stevin's results, and remained there until 1592. There are no contemporary documents reporting any experiment at the Leaning Tower, the only account we have being written by Galileo's last student Viviani, who wasn't even born until 1622. It is more likely that Galileo treated the equivalence of falling rates as an obvious common sense rule, and that the dropping of objects from "a high tower" in Pisa was more of a demonstration of this rule, than it was an experiment leading to its discovery.

It wasn't until Isaac Newton (1642–1727) that the rule was raised to the status of a fundamental principle of mechanics. Newton regarded it as such a cornerstone that he devoted the opening paragraph of his great 1687 treatise on mechanics *Philosophiae Naturalis Principia Mathematica* to it. To Newton, the principle of equivalence meant that the mass of any body, namely that property of a body (known as inertia) that regulates its response to an applied force, must be equal to its weight, namely that property that regulates the force exerted on it by gravity. Its consequence is that all bodies should fall in a gravitational field with the same acceleration, regardless of their structure or composition.

Newton did not stop there. He also elevated the principle to an experimental issue by trying to test it with good precision. If the principle of equivalence is correct, then the period of a pendulum of a given length should not depend on the mass or composition of the object suspended because the downward acceleration of the object is independent of its mass or composition. The period actually depends only on the length of the pendulum and on the common value of the gravitational acceleration. Newton suspended identical boxes of wood from wires 11 feet long. He filled one box with wood and the other with an equal weight of gold and set the pair of pendula in motion. He observed that the pendula kept in step to high accuracy. Repetitions of the experiment using silver, lead, glass, sand, salt, water, and wheat led to the same result, with the conclusion that the mass and weight of materials are the same, or that their accelerations are the same, to a precision of one part in a thousand. Further improvements by a factor of about 100 were made in the nineteenth and early twentieth centuries. However, pendulum experiments are limited in accuracy by several factors, including the effects of air currents set up by the swinging pendula, and the difficulty of timing moving objects accurately (imagine clocking a 3-second pendulum to a precision of hundredths or thousandths of a second).

It wasn't until the late nineteenth century that Roland von Eötvös (1848–1919) was able to achieve a significant improvement in accuracy, using an elegant new scheme. Eötvös was a distinguished physicist, having already made fundamental discoveries on the effect of temperature on the surface tension of liquids, the property that makes water drops on a table form rounded tops. In 1889, as a result of his achievements, he was elected president of the Hungarian Academy of Sciences. But by this time, he had turned most of his attention to the question of gravitational measurements.

Eötvös used a device known as a torsion balance, which he had originally developed to measure variations in the local acceleration of gravity for the purpose of geological studies. Two weights are attached to the ends of a rod, and the rod is suspended by a wire from a point that leads to a horizontal balance (the point need not be exactly in the center of the rod if the two

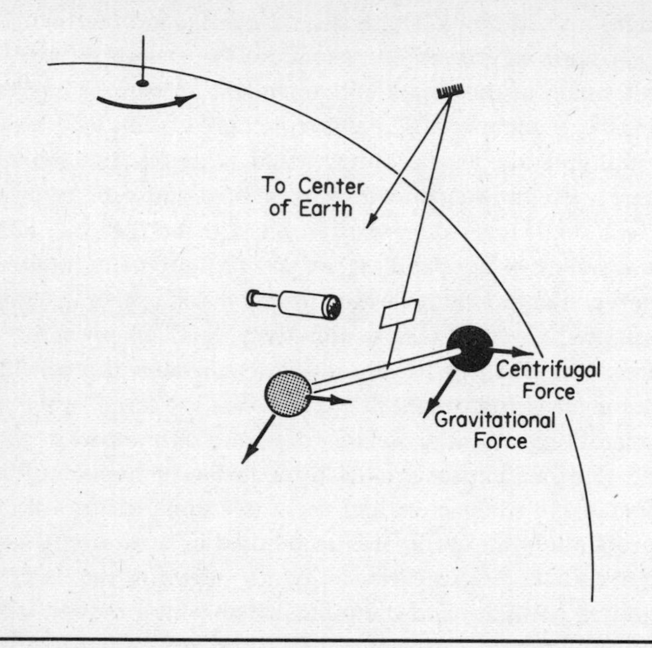

Figure 2.3 Eötvös experiment. Fiber supporting the rod does not hang exactly vertically because of the centrifugal force from Earth's rotation, so the downward gravitational force on the balls is not parallel to the fiber. If gravity pulls one material more strongly than the other, the rod will rotate about the fiber axis. If entire apparatus is rotated so that the two balls are interchanged, the resulting rotation will be in the opposite sense. The rotation is detected by observing light reflected from a mirror attached to the fiber.

weights are different). As it stands, this is an uninteresting setup because it merely measures the relative gravitational forces on the two objects. We also need an inertial force, like that exerted by the string on the pendulum in Newton's experiments. Fortunately, the Earth provides us with such a force automatically, as a result of its rotation. Unlike the gravitational force, which is directed toward the center of the Earth, the inertial centrifugal force is directed outward perpendicularly to the Earth's rotation axis (see figure 2.3). At the latitude of Budapest, where Eötvös's experiments were performed, the centrifugal force is about 400 times smaller than the gravitational force, and is directed about 47° from the vertical, toward the south. Suppose now that the two weights are made of different materials. If the inertial force acted differently in proportion to the gravitational force on one material compared to the other, there would be a resulting torque, or twisting force, that would cause a slight rotation of the rod about a vertical axis, until it was halted by the restoring torque of the twisted wire. The orientation of the rod relative to the laboratory can be measured in various ways. Of course, this doesn't tell us anything, because we don't know where the rod would have come to rest if the inertial force were turned off (unfortunately, we can't stop the rotation of the Earth with the flip of a switch). However, if the entire apparatus including the support for the wire is rotated by 180°, the torque would cause the rod to rotate in the opposite direction and come to rest in a different orientation. If there were no difference in the inertial effects on the different materials, there would be no torque, and thus no difference in the direction of the rod in the two configurations of the apparatus. This kind of experiment is called a "null" experiment, because the expected result, if the principle of equivalence is valid, is a null or zero difference between two measurements.

In two series of experiments, in 1889 and in 1908, Eötvös and his colleagues used platinum as one weight and different materials—copper, water, asbestos, aluminum, and so on—as the other. After correcting for the stray gravitational forces exerted on the apparatus by the experimenters themselves, they found no anomalous torques. Their conclusion was that the mass, or "inertial mass" and the weight, or "gravitational mass" of different materials is the same to a few parts in a billion. Another way of saying this is that different bodies suffer the same acceleration in a gravitational field to a few parts in a billion.

These experiments predated Einstein's work on gravitation, but, apparently, when he proposed his principle of equivalence as a foundation for gravitational theory in 1907, he was not aware of them. Instead, he accepted the equality of acceleration

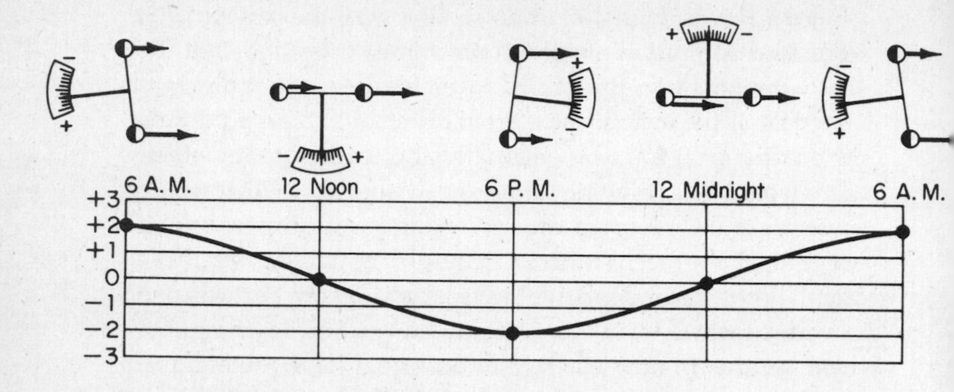

Figure 2.4 Princeton version of Eötvös experiment. Relative attraction of different bodies toward the Sun is measured. Sun is located to the right. If there is a difference between different materials, then as the rod's orientation relative to the Sun changes, the rod will swing in the manner shown, with a resulting signal of the deviation that varies on a 24-hour period. If there is no difference in attraction, the rod will not swing. Reprinted, by permission of the publisher, from "The Eötvös Experiment" by R. H. Dicke. Copyright © 1961 by Scientific American, Inc. All rights reserved.

as a given. By 1912, however, he had been apprised of Eötvös's experiments, and referred to them extensively in subsequent writings.

Eötvös's experiments represented the state of the art for almost sixty years. Finally, major improvements were made in the early 1960s and again around 1970 by two groups, the first at Princeton University (see figure 2.4), headed by Dicke, the second at Moscow State University, headed by Vladimir Braginsky. These groups had two things going for them. The first was a clever idea: Replace the Earth's gravitational force with that of the Sun, replace the centrifugal inertial force of the Earth's rotation with that due to the Earth's orbit around the Sun, and finally, let the Earth itself do the work of rotating the laboratory about the Earth-Sun line. Even though the gravitational force of the Sun on the bodies is about 1,000 times smaller, the angle between the direction of the supporting wire and the Sun is larger, and many sources of noise and error are eliminated by not having to rotate the apparatus bodily. The second thing these groups had going for them was the technology of the day, including excellent fibers for supporting the rod, good vacuum systems to eliminate the effects of air currents, automated temperature control systems, and sophisticated electrical and optical techniques for determining the orientation of the rod. If the equivalence principle were false, then as the Earth rotates, the rod should first turn in one direction when the Sun is "overhead," then, 12 hours later, turn in the other direction. The experimenters only had to look for rotations of the rod that varied on a 24-hour period. In neither experiment were any such variations found, to the limits of precision, quoted as 1 part in 100 billion by the Princeton group and as 1 part in a trillion by the Moscow group.* Therefore, bodies of different composition fall toward the Sun with the same acceleration to enormous precision.

One interesting and important conclusion can be drawn from these results. Because the mass of an atomic nucleus is made up of both the masses of the individual neutrons and protons, plus the mass equivalent of the internal energy due to the strong forces that bind the nucleus, and because different elements, such as aluminum and platinum, contain different amounts of nuclear internal energy per unit mass, then the energy of the nuclear forces must fall with the same acceleration as do the nuclear particles themselves. A similar conclusion applies to the electromagnetic energy associated with the forces between the charged protons and electrons. In chapter 7 we will ask whether or not gravitational energy also falls with the same acceleration.

The experiments that I have described employed macroscopic, laboratory-size bodies, and any conclusion that we make about the manner in which atomic particles or energies fall is necessarily somewhat indirect. What can we say directly

* For references to the original papers for these and other experiments described in this book, readers should consult *Theory and Experiment in Gravitational Physics* by C. M. Will (see suggestions for further reading at the end of this book).

about bodies of atomic dimensions? For example, do individual electrons fall with the same acceleration as do ordinary bodies? Unfortunately, this is an extremely difficult question to answer experimentally. Because the electron is charged, the slightest stray electric field in the apparatus will cause accelerations that swamp the gravitational effects being studied. For this reason the best Eötvös experiment on individual atomic particles was done using neutrons, which are electrically neutral. Although they are unstable outside the atomic nucleus, their long lifetime (1,000 seconds) permits numerous interesting observations. In one experiment performed in 1975 at the Technical University of Munich, slow neutrons from a nuclear reactor were emitted in a horizontal direction. After a distance of around 100 meters, the neutrons have fallen a measurable amount, and the small angle by which their motion deviates from the horizontal, about three ten-thousands of a degree, can be measured by reflecting them off a surface of a liquid mixture of lead and bismuth. To a precision of 1 part in 10,000, the angle agreed with what you would expect if the neutrons fell with the same acceleration as macroscopic bodies.

Because the principle of equivalence is so important, experimentalists are always looking for ways to improve the accuracy of its tests, or to find new ways to test it. Some of their ideas include using extremely low temperatures, near absolute zero, to reduce the errors caused by thermal fluctuations; and performing experiments in space, to reduce environmental sources of noise that affect any Earth-bound laboratory—for example, seismic disturbances (including trucks rolling by), and atmospheric effects, such as changes in barometric pressure. Perhaps some future space shuttle mission will find Sally Ride helping to set up an orbiting Eötvös experiment.

So far I have described how Einstein made the intuitive leap from the principle of equivalence to curved space-time, and I have described in some detail the experimental support for that principle, but I haven't said much about what curved space-time really is. Unfortunately, to do so would require a lot of mathematical detail that is beyond the scope of this book. Our goal here is to try to understand qualitatively what the observable consequences of curved space-time are, and how these consequences can be translated into real experiments.

Now, all the observable space-time effects that I will talk about in this book are extremely small, and many of them are rather subtle. But what about the gross, everyday effects of gravity, such as the tossing of a volleyball on Earth, or the orbit of the space shuttle about the Earth? I stated above that Einstein's proposal was that gravitation and curved space-time are in some sense one and the same thing. How on Earth (or anywhere else for that matter) is the motion of a volleyball related to curved space-time? Einstein postulated that the motion of a freely falling body, such as a thrown ball or an orbiting planet was along a geodesic, a "straight line" of the curved space-time. How can we reconcile that with what we observe to be the motion of a ball or planet, which certainly is not along anything approximating a straight line?

The reconciliation is really quite easy, once we have learned to distinguish space-time from space. The best way to do that is to draw a picture that includes some of the spatial dimensions, as well as the time dimension. We obviously can't draw all four space-time dimensions on a two-dimensional page, but if we ignore one of the spatial dimensions, then we can use perspective to represent the two spatial dimensions, plus time. These so-called "space-time diagrams" are a popular tool for talking about special and general relativity.

Let us first consider the volleyball problem (see figure 2.5). Imagine Sally Ride, who is also an excellent tennis and volleyball player, serving a volleyball high over the net, to a height of 10 meters, and coming down a distance 10 meters away. (Ordinarily, of course, Sally's service is a blistering bullet that must travel at half the speed of light, as I learned the hard way during lunchtime matches behind the Stanford gymnasium.) The trajectory of the volleyball is a parabola, certainly nothing like a straight line. However, let us consider the motion of the

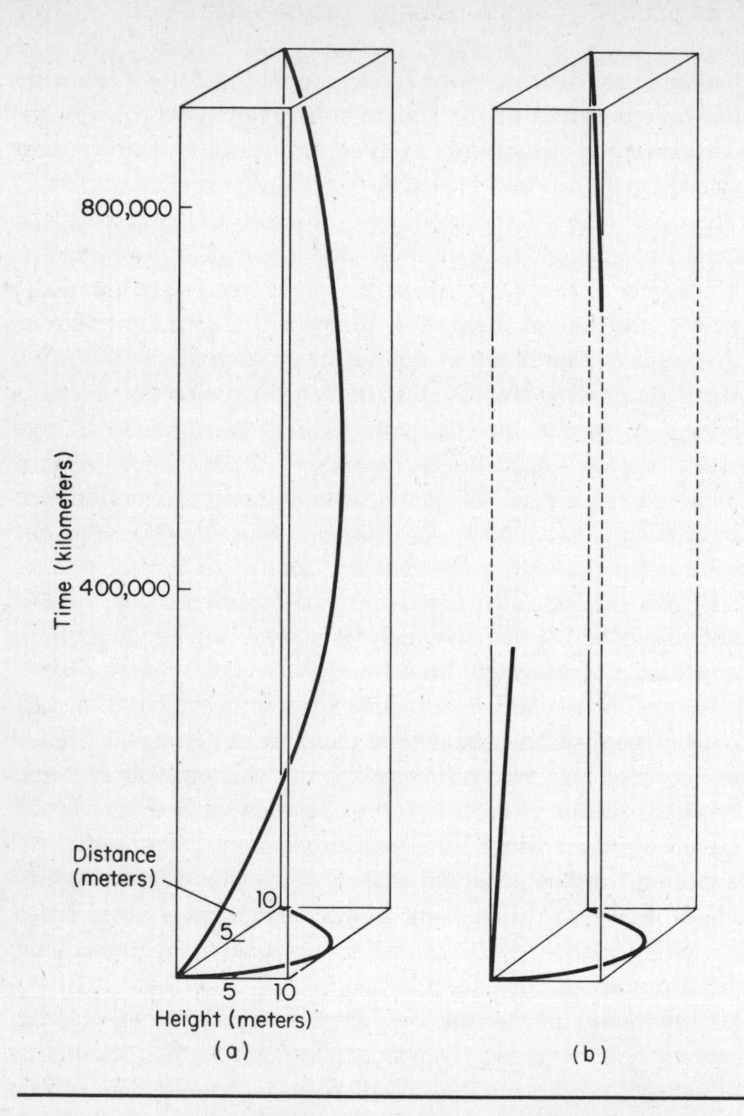

Figure 2.5 Space-time diagram for volleyball serve. Height is plotted to the right, distance along the ground into the page, and time upwards. In (a), the scale of the time axis has been severely compressed from 900,000 kilometers to the size of the page. Space-time path of volleyball runs from lower-left corner to upper-back corner. Projection of the path in space (a parabola) is shown at the base. If the time axis is stretched toward its true length, the space-time curve appears almost straight, as in (b).

ball in a space-time diagram. The base of the diagram shows a line that records the height of the ball (parallel to the page), and perpendicular to that (into the page) a line drawn between the starting and ending points of the ball to record its ground track. The third spatial dimension is not shown. The vertical line in the space-time diagram is the time direction. Now, because we know that we must treat space and time on an equal footing, we must therefore give them the same units. But how do we do this if spatial distances are measured in meters, say, and time is measured in seconds? The way to do it is through the speed of light, which according to special relativity is a universal constant, the same value when measured in any inertial frame. Therefore, if we take a time interval of 1 second and multiply it by the speed of light, approximately 300 million meters per second, then we have something with units of length. This length is just the distance traveled by light in 1 second (300,000 kilometers, or 186,000 miles). If we now look at time using units of distance, we would say that "1 meter of time" corresponds to the time it takes light to travel 1 meter, or about 3.3 nanoseconds (1 nanosecond equals one-billionth of a second). We are told to treat time and space equally, so if we mark off distances on the spatial lines of our space-time diagram in intervals of 1 meter, we must also, on the same scale, mark off intervals of 1 meter on the time line. Let us now draw the trajectory of the volleyball on the space-time diagram, each point being determined by the horizontal distance from Sally, the height, and the corresponding time. Right away we have trouble. The time taken for the ball to reach its highest point is only 1.4 seconds, but on the time line, this corresponds to 430,000 kilometers, or a bit farther than the distance to the Moon. And that is only one-half of the trajectory of the ball! Clearly, our space-time diagram will not fit on one page of this book. Nevertheless, we begin to see the sense in which the trajectory of the volleyball is a geodesic or "straight line." The trajectory begins at the starting point; then as we move up the time line, it moves into the page along the ground track line as well as to the right as it ascends. Continuing up the time line (we are now past the distance to the Moon), we find that the point stops moving to the right (the high point of the ball's motion) and begins to move to the left, while continuing along the ground track line. Finally, at the top of the space-time diagram, the point reaches the end of the trajectory, on the original line. It is clear that the trajectory has curved, but because it has been stretched in the time direction to more than twice the distance to the Moon, we would be hard pressed to look at any piece of the curve and say that it was anything but a straight line. The point is that when we look at the trajectory of the ball in space, it describes a parabola, but when we look at it in space-time it is almost, though not quite, a straight line. The fact that it is nearly straight in space-time is a consequence of the smallness of space-time curvature on Earth.

Another illustration of this is the orbit of the space shuttle around the Earth (see figure 2.6). In space, the orbit is nearly a circle. But in space-time, the length of time line required to incorporate one orbit of the shuttle, about an hour and a half in time, would be about 1.6 billion kilometers, or beyond the orbit of Saturn. In space-time, the shuttle's orbit would be a helix, but so stretched in the time direction that it too would not be much different from a straight line.

In the language of curved space-time, the solar system is said to be a "weak field," or "low curvature" system. The space-time trajectories of freely falling bodies are geodesics that deviate only slightly from ordinary straight lines. As a consequence, the effects of curved space-time that go beyond the gross features of these trajectories are just tiny corrections and are therefore difficult to detect and measure. This is what has made experimental gravitation such a challenge for seventy years.

One final remark. Nowhere in this chapter have the words "general relativity" appeared, until now. Everything that I have discussed is a direct consequence of the principle of equivalence and of Einstein's leap to curved space-time. This is such a fun-

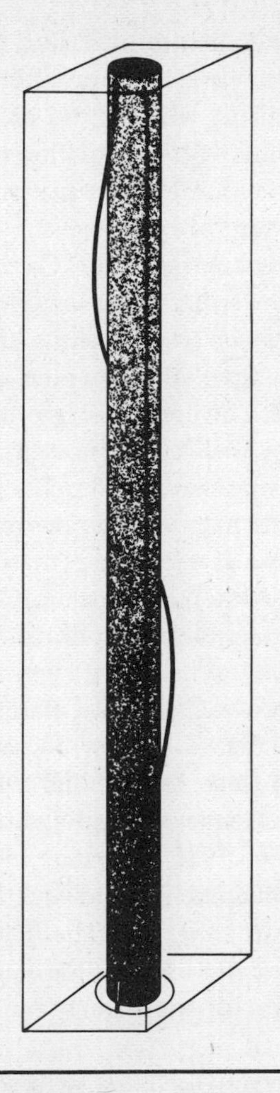

Figure 2.6 Space-time diagram for space shuttle. Time axis has again been compressed. Shaded tube is the path of the Earth's body through space-time. Path of the shuttle is a helix in space-time, a circle in space.

damental idea that we now regard it as totally separate from general relativity, although in Einstein's mind, the two were inextricably tied. The Eötvös experiment is viewed as one of the primary tests of the validity of curved space-time; it is such a basic experiment that much effort has gone into carrying it out, and much more effort will go into improving it as much as possible in the future. Were a violation of the equality of acceleration to appear at some level, it would require a complete revision of our picture of space-time.

What, then, is general relativity? General relativity is really a separate theoretical entity. It assumes the validity of curved space-time, but goes on to answer the kind of question that we have avoided asking in this chapter, namely, how much is space-time curved? The equivalence principle only tells us that it is curved; it can't tell us by how much. General relativity provides a set of mathematical equations, called "field equations" that allow us to calculate how much space-time curvature is generated by a given lump of matter, such as the Sun, the Earth, or a stone. Although Einstein had a clear physical understanding of the equivalence principle and of curved space-time as early as 1907, it took him eight more years to arrive at the equations of general relativity. Part of this time was spent learning the mathematics of curved space-time, and part of it was lost following leads down blind theoretical alleys, but finally, during an intense and exhausting three-week period during November of 1915, he made his second great leap of imagination and obtained the field equations in their final form. The theory was now complete. The field equations determine how much curvature there is, and the equivalence principle tells us how matter responds to it: Freely falling bodies move along geodesics.

The principle of equivalence is so powerful and so well verified by the Eötvös experiment that most modern theories that have been proposed as alternatives to general relativity are also based on curved space-time, in exactly the same way. Such theories would be in agreement with every statement made in this chapter. They differ from general relativity only in the field equations that they provide for determining how much curvature there is. The Brans-Dicke theory is one of these theories.

Experimental tests of gravitational theories can therefore be divided into two classes: those that test the principle of equivalence, such as the Eötvös experiment, and those that test specific gravitational theories. Before we talk about the latter class of tests, we must discuss an experiment that Einstein devised as a test of general relativity, but that we now realize is actually another test of the principle of equivalence, and thus is just as fundamental as the Eötvös experiment. That test is known as the gravitational red shift.

3

The Gravitational Red Shift of Light and Clocks

BOB VESSOT had a right to be a little nervous as he looked up at the launch gantry. There, perched at the top of the 73-foot-tall Scout D rocket and hidden behind a cover that looked like the cap on a ball-point pen, was one of the most precise atomic clocks ever constructed. It was an instrument so delicate that it could maintain its rate to within a trillionth of a second each hour. It represented five years of his life, a masterful effort by his laboratory team, and several million dollars of NASA investment.

The clock had never flown on a rocket before. The only reason for thinking it might survive the 20 g's of thrust, or 20 times the force of gravity, it would experience on takeoff was the extensive series of shake and centrifuge tests it had successfully passed at the Marshall Space Flight Center in Alabama before being brought here to NASA's launch facility on Wallops Island, one of the many small islands that hug the eastern coast of the narrow Virginia peninsula separating Chesapeake Bay from the Atlantic Ocean.

Vessot shuddered at the possibility that the rocket might go off course. In that event, the flight controllers had orders to blow it up. During the weeks of getting things ready for the flight, Vessot had gotten the impression that the Wallops personnel were rusty. Because of a recent dry spell in suborbital missions, they hadn't carried out a launch in a long time. One countdown had already been aborted because of a problem with some ammonia refrigerant. Still, the Scout D was one of the most reliable rockets around.

There was only one other clock in the world as good as the one on the rocket, and that was its twin; it was currently located 720 miles to the south, at the NASA tracking station at Merritt Island, right next to Cape Canaveral. During the two-hour suborbital flight of the Scout D, the rates of the two clocks were to be compared with each other. If all went well, the experiment would be able to see, to higher precision than ever before, if Einstein was right about the effect of gravity on clock rates, the effect known as the gravitational red shift.

It was a perfect day for a launch, a pleasant June morning in 1976, with just some high, thin clouds in the sky. Following the flight, the data would be assembled and taken back to Harvard University. After the tension of the last few months, Vessot looked forward to the long July 4th weekend—perhaps a relaxing sail on the ocean near his Marblehead home, or a look at the U.S. bicentennial celebrations—then it would be back to work at his Harvard laboratory and the painstaking analysis of the data.

As the second countdown proceeded, Vessot might also have looked backward to the origins of this effect, to whose measurement he had devoted so much effort. The gravitational red shift was the first of Einstein's great predictions, made as soon as he had formulated his version of the principle of equivalence and recognized what it meant for the nature of physics in freely falling laboratories. This was 1907, eight years before the full formulation of general relativity. Yet, surprisingly, it was the last of Einstein's three predictions to be tested with any degree of accuracy.

The easiest way to understand the gravitational red shift is to consider a simple thought experiment involving the Earth, an emitter and receiver of light, and some freely falling laboratories. This thought experiment is actually very similar in concept to the first real experiment to measure the red shift accurately. Imagine an emitter of a well-defined frequency or wavelength of light placed at the top of a high tower with its beam directed downward. A tunable receiver is placed on the ground and is tuned to receive the incoming signal from the emitter on the tower. Compared to the emitted frequency, is the received frequency larger, smaller, or the same? Let us use the principle of equivalence of Einstein to answer this question.

We want to focus on a small portion or "packet" of the emitted light.* Let us imagine a laboratory that was earlier shot out from the center of the Earth, but had its rocket boosters turned off, so that it is now in free fall, even though it is still rising away from the Earth (see figures 3.1 and 3.2). Let us suppose that the timing of its takeoff and its starting velocity were cunningly chosen so that at the exact instant our wave packet is emitted by the emitter, the laboratory has reached the top, or apogee, of its trajectory, and is sitting momentarily at rest right next to the emitter, just about to begin to fall back to Earth. Because this is a thought experiment, we don't need to worry about how to achieve such an incredible feat. Such de-

* According to quantum mechanics, such a packet could correspond to a fundamental "particle" or quantum of light called the photon; or, according to classical electromagnetism, it could correspond simply to a short stretch of a continuous wave. This "wave-particle" duality is a basic feature of nature, and whether light needs to be described as a particle or as a wave depends on the experiment being performed. However, the principle of equivalence does not distinguish between the two natures of light. We would get the same answer no matter which description of light we used. This is why we don't really need to specify the nature of our packet of light precisely.

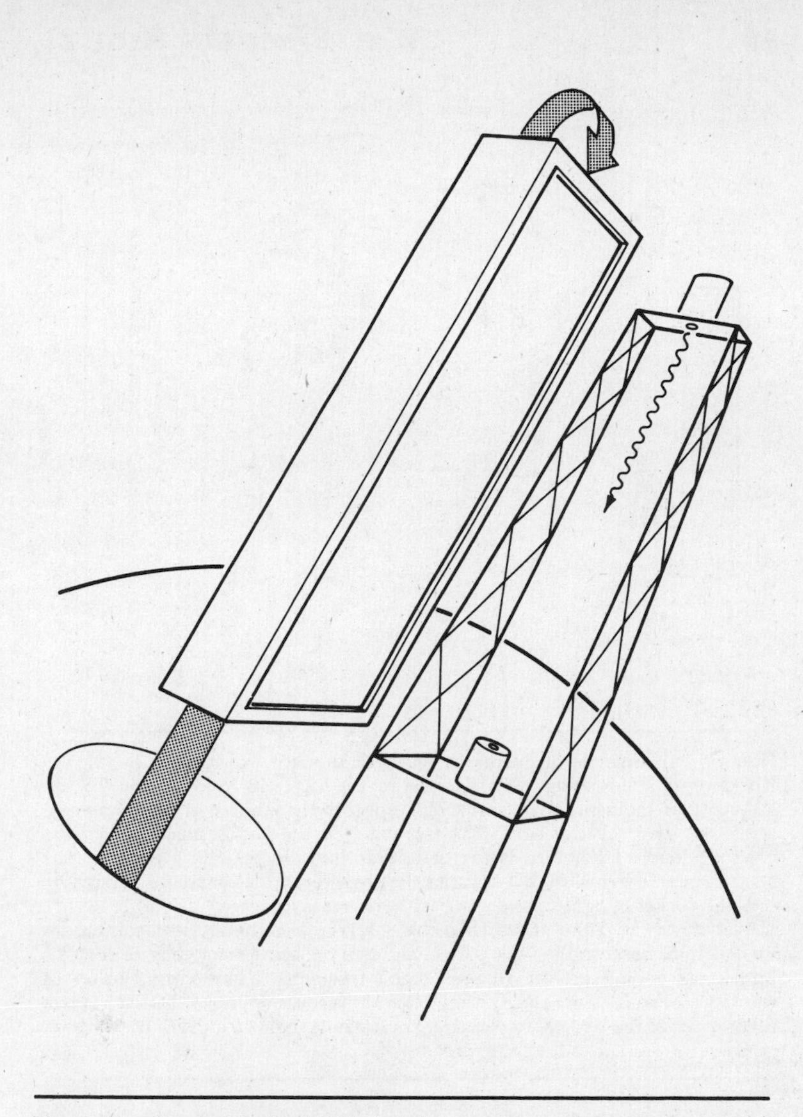

Figure 3.1 Gravitational red-shift thought experiment. Emitter at the top of a tower sends a packet of light toward receiver at the bottom. A laboratory was earlier shot out from the center of the Earth and had its rockets turned off so that it is in free fall (no gravity inside) thereafter. It reaches the top of its trajectory at the moment the emitter sends the packet of light. The observer inside can use the gravity-free laws of special relativity to analyze the emission, propagation, and reception of the packet.

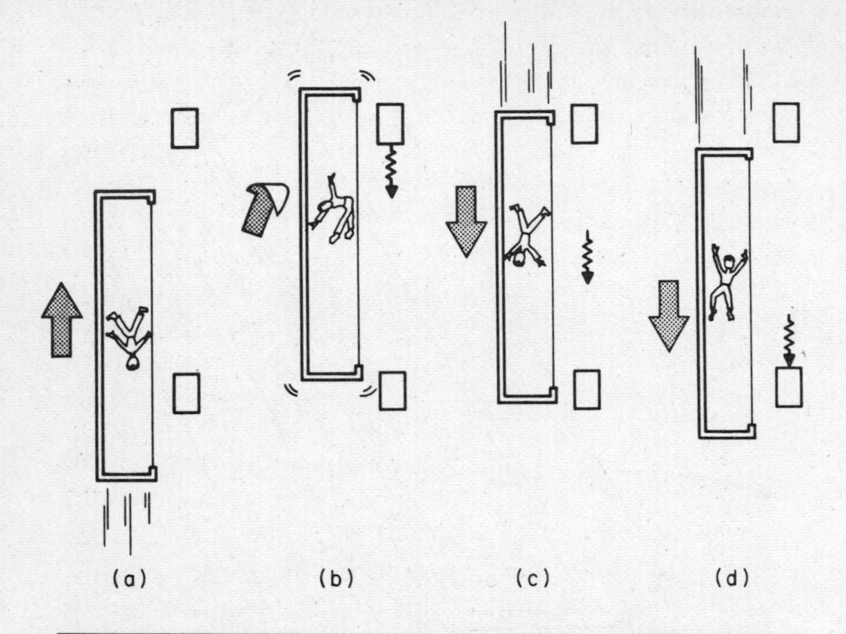

Figure 3.2 Sequence of events in red-shift thought experiment. In (a), the laboratory is still moving upwards, nearing the top of its trajectory. In (b), the laboratory is momentarily at rest with respect to the emitter, at the moment it sends out the packet of light. The frequency of the packet measured by the observer in the laboratory is the standard value. In (c), the laboratory has started to fall downward, but because the observer inside senses no gravity, he sees the packet of light propagating with the same frequency as before. In (d), the laboratory is falling faster, and the observer sees the receiver coming up toward him. Because the light packet still has the same frequency as seen by him, the ascending receiver will see a higher frequency (a blue shift), because of the Doppler shift. The velocity that determines the amount of the shift is just the speed that the laboratory picked up in the time it took for the light packet to go from the emitter to the receiver.

tails are irrelevant to the experiment at hand anyway. The important thing is what happens now.

An observer inside the laboratory measures the frequency of the emitted packet of light. Because he is in free fall, and because he is well versed in the principle of equivalence, he realizes that from his point of view gravity is absent, and so the emitted frequency can be calculated using the laws of special relativity. This, of course, is why it was important to make his laboratory freely falling. Also, because the laboratory is at rest with respect to the emitter, if only for a moment, the emitted frequency is the "rest" frequency, unaffected by any slowing down of moving clocks that is predicted by special relativity. The measured frequency would therefore be the standard value for that emitter, and could be looked up, say, in standard tables of physical constants, or calculated using the standard laws of atomic or nuclear physics.

The observer then follows the progress of the wave packet as it travels toward the bottom of the laboratory. Because he is still in free fall, the motion of the wave continues to obey the laws of special relativity, including moving at the constant speed of light with an unchanging frequency. But as the wave packet moves downward, the observer's laboratory also begins to fall. It was at its apogee only for a moment; then the gravitational force of the Earth began to pull it back down. Nevertheless, the frequency of the wave packet remains unchanged as seen by him. After a moment, the observer notices the receiver coming up toward him, because he is falling, while the receiver is at rest on the surface of the Earth. Thus, when the onrushing receiver receives the packet of light, it is going to detect a higher frequency than that measured in the freely falling laboratory, because of the Doppler effect. Here, of course, the emitter and receiver are still at rest with respect to each other; the important point is that from the point of view of the observer in the freely falling laboratory, in which the frequency has its standard value, the receiver is moving toward him. The velocity of the laboratory relative to the receiver is the same as the velocity that the freely falling laboratory has picked up in the time taken for the wave packet to travel the distance between the emitter and the receiver. This is given by the acceleration of gravity (980 centimeters per second per second) times the time, which is the separation between the emitter and the receiver divided by the speed of light. The fractional shift in frequency is then given by this velocity divided by the speed of light. For example, for a difference in height of 100 meters, the shift would be only ten parts in a million billion, or one-trillionth of a percent. If the emitter and receiver are at the same height, but separated in the horizontal direction, there is no frequency shift.

In this thought experiment, the observed shift was toward higher frequencies—the blue end of the visible spectrum—because the freely falling frame was heading toward the receiver. If the emitter had been at the bottom and the receiver at the top, the shift would have been toward lower frequencies—the red end—because by the time the wave packet reached the top, the freely falling frame would be falling away from the receiver. Even though the result can be either a red shift or a blue shift, depending on the experiment, the generic name for this effect is the gravitational red shift. It is called a "gravitational" shift because it occurs only in the presence of a mass that exerts a gravitational force.

It should be apparent from our thought experiment that the gravitational red shift is a truly universal phenomenon. It was the behavior of the freely falling laboratory that was the crucial element in the analysis. The nature of the emitter and receiver did not play a significant role, nor did our treatment of the nature of light. The light could have been in the visible spectrum, or it could have been in the radio or in X rays. The signal could have been a continuous beam, or it could have been in the form of beeps, such as might be emitted by a strobe light set to flash once per second. In the latter example, the observer at the bottom of the tower would observe not only that the intrinsic frequency of the light emitted by the strobe was shifted toward the blue, but also that the flashes arrived more quickly than once per second. Thus, all frequencies appear to be shifted. If the strobe's flashes were timed by some sort of clock, then the observer on the ground would argue that the clock at the top of the tower was ticking faster than his ground clock, in other words, that the clock rate was "blue shifted." In fact, the distinction between clock and emitter or receiver of light that we have used is purely a semantic one. The term "clock" really means a device that performs some physical activity repetitively at a well-defined, constant rate. The activity could be the mechanical sweep of a second hand, the flashes of a strobe, or the waves of an electromagnetic signal. Modern atomic clocks are based on the latter phenomenon—the emission of light with a constant, stable, well-defined frequency. The gravitational red shift affects all clock rates equally, whether mechanical, biological, or atomic. As we will see, atomic clocks provide the best test of the gravitational red shift.

Another thing that should be apparent is that we again did not use general relativity anywhere in the discussion. The gravitational red shift depends only on the principle of equivalence. Even though Einstein viewed the red shift as one of his three main tests of general relativity, we now regard it as a more basic test of the existence of curved space-time. Any theory of gravity that is compatible with the equivalence principle (and there are many, including, for instance, the Brans-Dicke theory) automatically predicts the same gravitational red shift as general relativity.

A question that is often asked is, Do the intrinsic rates of the emitter and receiver or of the clocks change, or is it the light signal that changes frequency during its flight? The answer is that it doesn't matter. Both descriptions are physically equivalent. Put differently, there is no operational way to distinguish between the two descriptions. Suppose that we tried to check whether the emitter and the receiver agreed in their rates by bringing the emitter down from the tower and setting it beside the receiver. We would find that indeed they agree. Similarly, if we were to transport the receiver to the top of the tower and set it beside the emitter, we would find that they also agree. But to get a gravitational red shift, we must separate the clocks in height; therefore, we must connect them by a signal that traverses the distance between them. But this makes it impossible to determine unambiguously whether the shift is due to the clocks or to the signal. The observable phenomenon is unambiguous: the received signal is blue shifted. To ask for more is to ask questions without observational meaning. This is a key aspect of relativity, indeed of much of modern physics: we focus only on observable, operationally defined quantities, and avoid unanswerable questions.

There is, however, one way to see the effect of the gravitational red shift without an intervening signal, and that is to measure its effect on the elapsed time of two clocks. Begin with two clocks side by side, ticking at the same rate, and synchronized, so that at some chosen moment they read the same time. Take one clock slowly to the top of the tower and let it sit there for a while. Then bring it back down slowly and compare it with the ground clock. While their rates will once again be the same, the tower clock will be ahead of the ground clock. The inference from this is that the tower clock ran faster while it was on the tower, but unless we connect the clocks by a light signal, we cannot see the rate difference except after the fact. This idea actually was the basis for an interesting experiment using atomic clocks and jet aircraft, to be described shortly.

Early attempts to measure the gravitational red shift focused on light from the Sun and from white-dwarf stars. When an atom undergoes a transition from one electronic level to another, it emits light at a frequency or wavelength that is a characteristic of the atom. In the laboratory, the frequencies of these "spectral lines" can be measured with high accuracy. The same atom on the surface of the Sun will emit light whose frequency is red shifted as seen from Earth because, in a thought experiment using a tower sitting on the surface of the Sun and stretching all the way to the Earth, the atom would be at the bottom of the tower and the receiver on Earth would be at the top of the tower. Here, the problem is complicated slightly by the fact that the gravitational acceleration is not the same over the whole length of the tower, varying from a maximum value at the bottom, that is, on the solar surface, to a very small value at the top, that is, at the orbit of the Earth. In our derivation of the gravitational red shift using freely falling laboratories, we assumed implicitly that the acceleration was the same throughout. This complication only means that the effect is expressed not in terms of the acceleration and the height difference, but instead in terms of the difference in the values of the gravitational potential between the emitter and the receiver. The gravitational potential is the quantity whose variation in space determines the gravitational acceleration. For the red shift of solar spectral lines, the effect is given by the difference between the potential at the surface of the Sun and that at the surface of the Earth. Because the gravitational potential on the Earth's surface is 3,000 times smaller than that on the Sun, we can ignore it; we can also ignore the motion of the Earth in its orbit. These produce tiny corrections in the frequency, too small to be observed in this case. For a wavelength of 5,893 Angstroms (an Angstrom is ten billionths of a centimeter), corresponding to the bright-yellow emission line of the sodium atom, one of the most intense in the solar spectrum, the shift is 0.0125 Angstroms toward longer wavelengths (lower frequencies), well within reach of standard techniques.

Unfortunately, attempts to measure the solar shift in this and other atomic lines between the years 1927 and 1960 failed to agree with the predicted value. This was regarded not as a failure of the equivalence principle's prediction but rather as a failure in our understanding of the solar surface. The gas at the solar surface experiences violent motions, with rising columns of hot gas and falling columns of cooler gas, which lead to Doppler shifts of the emitted frequencies both to the blue and to the red. It is also under high pressure, which causes shifts to the blue and red in the intrinsic frequencies emitted by the atoms themselves. These and other effects made it impossible to observe the gravitational shift clearly. It wasn't until the 1960s and 1970s, when these effects were better understood, that astronomers were able to measure the gravitational red shift of solar lines. The results agreed with the prediction to about 5 percent.

Similarly unsuccessful were attempts to detect the red shifts of lines from white-dwarf stars, such as the companion of Sirius, the dog star. White dwarfs are stars with masses comparable to that of the Sun, but with diameters of only thousands of kilometers, or 100 times smaller than the Sun. The gravitational potential at their surface is 100 times larger than the Sun's, and consequently the red shift is 100 times larger, and easier to detect. Unfortunately, we need to know the mass and radius of the white dwarf with some precision in order to make a prediction of the red shift against which to compare the observations. In the case of the companion of Sirius, for example, the mass is about the same as that of the Sun, and the radius is about 40 times smaller, but a host of complicated effects, too numerous to describe, make it impossible to formulate mass and radius estimates more accurate than this. This test of the gravitational red shift remains unfulfilled to this day.

The first truly accurate test of the red shift, and the experiment that helped usher in the new era for general relativity, was the Pound-Rebka experiment of 1960. This experiment is very close in concept to the one described in our thought experiment in figure 3.2. In this case, the tower was the Jefferson Tower of the physics building at Harvard University. For the tower's height of 74 feet the predicted frequency shift is only two parts in a thousand trillion; an emitter and receiver of extremely well-defined frequency is required. One possibility was the nucleus of the unstable isotope of iron, denoted Fe^{57}, with a lifetime or half-life of one ten-millionth of a second (one-tenth of a microsecond). When this isotope decays, it emits light in the form of gamma rays of energy 14,400 electron volts, or wavelength 0.86 Angstroms, within a very narrow range in wavelengths of only one part in a trillion of the basic wavelength. The same isotope can also absorb gamma rays of the same frequency within the same narrow spread. However, this alone is not enough to measure the red shift. Inside any realistic sample containing Fe^{57}, the iron nuclei are constantly in motion because of heat energy. This leads to Doppler shifts in the emitted gamma ray frequencies, whose result is to broaden the range of wavelengths. In addition, upon emission or reception of a gamma ray, the iron nucleus recoils, and this recoil velocity also causes a Doppler shift in the frequency. These effects can broaden the Fe^{57} range of frequencies so severely that a red-shift measurement would have been impossible, had it not been for Rudolph Mossbauer. Working at the Max Planck Institute in Heidelberg, Germany, in the late 1950s, Mossbauer discovered that if such a nucleus were implanted in the right kind of crystal, then the forces of the surrounding atoms not only reduce the heat-induced oscillations of the atom but also transfer the recoil momentum of the emitting atom to the crystal as a whole, thereby virtually eliminating its recoil velocity. For this discovery, Mossbauer was awarded the Nobel Prize in Physics in 1961. The Harvard gravitational red-shift experiment was one of the many important applications of the effect cited at the awards ceremony by the Swedish Academy of Sciences.

Pound and Rebka used the Mossbauer effect in their experiment in order to prepare emitters and receivers (absorbers) of gamma rays with an extremely narrow range of frequencies. Still, the range of frequencies of Fe^{57} was 1,000 times larger than the size of the expected shift, so they had to measure the shift of the range of frequencies from the emitter relative to the range of frequencies that could be absorbed by the receiver to better than one part in a thousand of that range. To do this, they put the emitter on a movable platform that could be raised and lowered slowly using a hydraulic lift and a rack-and-pinion clock drive. If the emitter was at the top of the tower, so that the gamma rays would be blue shifted upon reaching the bottom, the platform was raised slowly, producing a Doppler shift toward the red. By adjusting the rate at which the emitter was raised, Pound and Rebka could produce a Doppler red shift that would cancel or compensate for the gravitational blue shift, thereby allowing the range of frequencies of gamma rays received at the bottom to match closely the range that could be absorbed by the receiver. The Doppler shift required to do this was then a measure of the gravitational blue shift. The needed velocity was about 2 millimeters per hour. In order to eliminate certain sources of error, the actual experiment was a symmetrical one. Half the measurements were made with an emitter at the top and an absorber at the bottom, to measure the blue shift, and half were made with an emitter at the bottom and an absorber at the top, to measure the equal and opposite red shift. The results of the 1960 experiment agreed with the prediction to 10 percent, and those of an improved 1965 experiment version by Pound and Joseph L. Snider agreed to 1 percent.

Another way to check the gravitational red shift is to compare the readings of two clocks that have undergone a temporary separation. This idea is similar to the so-called twin paradox in special relativity, in which two twins (or identical clocks) are separated, one taking a journey to a distant galaxy and returning to find himself younger than his stay-at-home twin. According to special relativity, this is no paradox, the result being completely understood as a consequence of time dilation, or the slowing down of the traveler's clock relative to that of the twin. During October 1971, a remarkable experiment was performed that checked both these phenomena—gravitational red shift and time dilation—in their effects on traveling clocks. The idea behind the "jet-lagged clocks" experiment is this (see figure 3.3). Consider, for simplicity's sake, a clock on Earth at the equator, and an identical clock on a jet plane flying overhead to the east at some altitude. Because of the gravitational blue shift, the flying clock will tick faster than the ground clock. What about the time dilation? Here we must be a bit careful, because both clocks are moving in circles around the Earth rather than in straight lines. Now, according to special relativity, the rate of a moving clock must always be compared to a set of clocks that are in an inertial frame, in other words that are at rest or moving in straight lines at constant velocity. Therefore we can't simply compare the flying clock directly with the ground clock. Let us instead compare the rates of both clocks to a set of fictitious clocks that are at rest with respect to the center of the Earth. The ground clock is moving at a speed determined by the rotation

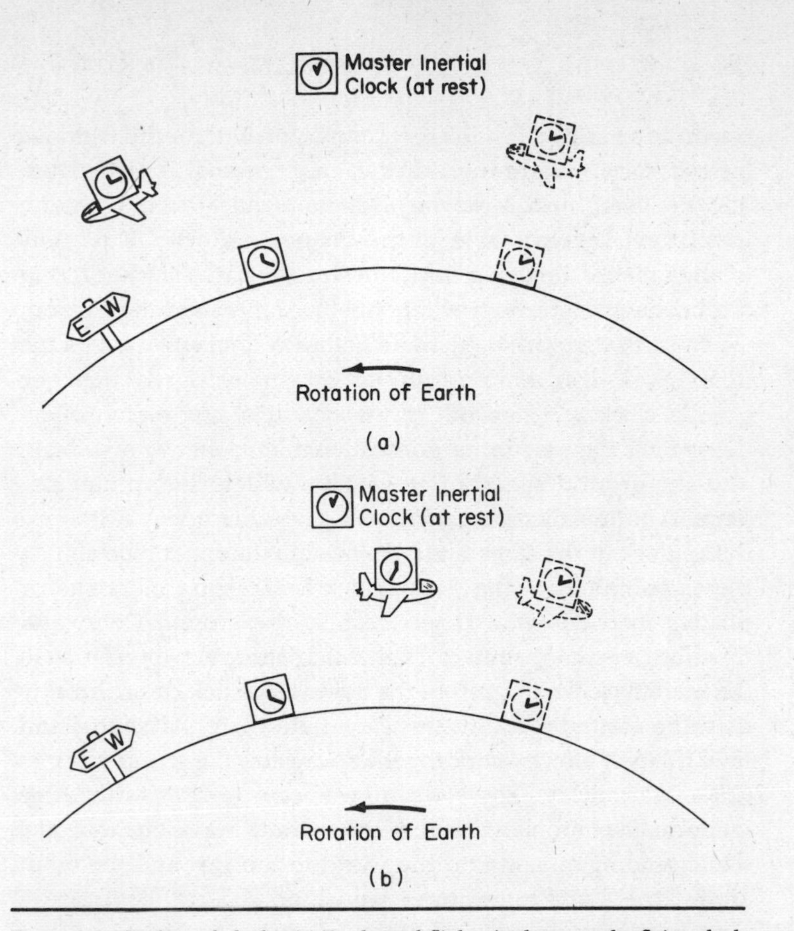

Figure 3.3 Jet-lagged clocks. (a) Eastbound flight. At the start, the flying clock is directly above the ground clock; after some time it is to the east. The flying clock travels more quickly relative to a stationary fictitious master clock than does the ground clock. Thus, the flying clock ticks more slowly relative to it than does the ground clock because of the time dilation of special relativity, so the flying clock also ticks more slowly than the ground clock. On the other hand, the gravitational blue shift makes the flying clock tick more quickly than the ground clock. The two effects can offset each other, and whether the net effect is a gain or loss of time by the flying clock depends on the speed and altitude of the flight. (b) Westbound flight. At the start, the flying clock is directly above the ground clock; after some time it is to the west. But because of the Earth's rotation, it has moved to the east relative to the inertial clock, with a lower velocity than has the ground clock. Thus, the ground clock ticks more slowly relative to the master clock than does the flying clock, so the flying clock ticks more quickly than the ground clock as a result of time dilation. The gravitational blue shift also causes the flying clock to tick more quickly, so the two effects augment each other.

rate of the Earth, and thus ticks more slowly than the fictitious inertial clocks (as represented by a master inertial clock in figure 3.3); the flying clock is moving even more quickly relative to the inertial clocks, so it is ticking even more slowly. Thus time dilation makes the flying clock run slowly relative to the ground clock. The two effects, gravitational blue shift and time dilation, tend to offset one another, and whether the net effect is that the flying clock ticks more quickly or ticks more slowly than the ground clock will depend on the height of the flight, which determines the amount of gravitational blue shift or speed-up, and the ground speed of the flight, which determines the amount of time dilation slow-down. Consider now a westward flying clock at the same altitude. The gravitational blue shift is the same, but now the flying clock is traveling more slowly relative to the inertial clocks than is the ground clock, and therefore it is the ground clock that ticks more slowly relative to the inertial clocks. Therefore the flying clock ticks more quickly than the ground clock. In this case, both the gravitational and time dilation effects work together, causing the flying clock to tick more quickly. Therefore, if we were to start with three identical, synchronized clocks, and were to leave one at home while sending one around the world to the east and the other around the world to the west, we would expect the westward clock to return having gained time, or aged more quickly, while the eastward clock would have gained or lost time depending on the altitude and speed of the flight.

The actual experiment, coordinated by J. C. Hafele, then of Washington University in St. Louis, and Richard Keating of the U.S. Naval Observatory, used cesium-beam atomic clocks. Because of the expense involved, they could not simply charter planes to circumnavigate the globe nonstop, but instead they had to fly the clocks on commercial aircraft during regularly scheduled flights. (Because of government regulations, they couldn't even fly first class!) No, the clocks were not strapped into their seats like the other passengers. Actually on most of the flights they were positioned against the front wall of the coach-class cabin to protect them against sudden motions on landing and to connect them more easily to the airplane's power supply. The flights included numerous stopovers during the course of the experiment, and the air speeds, altitudes, latitudes, and flight directions all varied. But by keeping careful logs of the flight data, they could calculate the expected time differences for each flight. The eastward trip took place between October 4 and 7 and included 41 hours in flight, while the westward trip took place between October 13 and 17, and included 49 hours in flight. For the westward flight the predicted gain in the flying clock was 275 nanoseconds (billionths of a second), of which two-thirds was due to gravitational blue shift; the observed gain was 273 nanoseconds. For the eastward flight, the time dilation was predicted to give a loss larger than the gain due to the gravitational blue shift, the net being a loss of 40 nanoseconds; the observed loss was 59 nanoseconds. Within the experimental errors of ±20 nanoseconds, attributed to inaccuracies in the flight data and intrinsic variations in the rates of the cesium clocks, the observations agreed with the predictions.

The Pound-Rebka-Snider series of experiments that I described earlier was a tour de force of laboratory technique. It gave the first verification of the gravitational red shift, and, as we see in hindsight, it foreshadowed a new era of experimental gravitation. It was so beautifully done that it reached the ultimate limit for that technique of using gamma-ray emitters and the Mossbauer effect. In twenty years, it has never been improved upon. Although an accuracy of 1 percent is impressive, the red shift is such a fundamental effect that one would like to do better. The Hafele-Keating jet-lagged clocks experiment did not achieve the same kind of accuracy as the Pound experiments, but it did verify the cumulative effect of the red shift and time dilation for atomic clocks. The question was, Could the red shift be measured more accurately?

In fact, by the time of the jet-lagged clocks experiment, plans were well under way to try. What is the main ingredient missing from the two experiments I have just described? Height. Up to a limiting value, the size of the gravitational red shift increases with the difference in height between the emitter and receiver or between the two clocks. The gamma-ray experiment unfortunately could not go to larger differences in height, because the gamma rays are emitted equally in all directions from the sample of Fe^{57}. A consequence of this is that as the height is increased, the number of gamma rays received by the absorber becomes so small as to be unusable. The jet-lagged clocks experiment was financially limited to typical commercial aircraft altitudes, but it contained the elements of the ultimate experiment, namely, putting an atomic clock on a satellite or rocket.

This idea had been suggested as early as 1956, just before the first Earth satellites were launched, and had been tried in 1966, with modest success, at the 10-percent level. But the experiment that was being worked on in 1971 was truly ambitious. The idea was to get two of the best atomic clocks in existence, called hydrogen maser clocks, put one on the top of a rocket, blast it up to a couple of times the radius of the Earth, and compare its rate to the clock left on the ground. In principle, the accuracy achievable in a measurement of the shift was 1 part in 10,000, or one-hundredth of a percent. Fortunately, the experiment brought together the two sets of experts required to pull it off.

The first set of experts consisted of Robert Vessot and Martin Levine of the Smithsonian Astrophysical Observatory at Harvard University. Their laboratory was at the forefront of development of this new kind of atomic clock. Soon after the invention of the hydrogen maser clock in 1959 by Harvard physicists Norman Ramsey, Daniel Kleppner, and H. Mark Goldenberg, Vessot, who then worked for Varian Associates, pioneered the development of a commercial, portable version of the new timepiece. By 1969, Vessot had left industry for the halls of Harvard, and now wanted to make use of these devices in fundamental physics experiments. The other set of experts consisted of the National Aeronautics and Space Administration, which would provide the launch vehicle, tracking, and other facilities required to get the clock aloft and measure its frequency shift.

The hydrogen maser clock was ideal for this experiment. It is based on a jump of the electron (called a "transition") between two atomic energy levels in hydrogen that emits light in the radio portion of the spectrum, with a frequency of 1,420 million cycles per second (1,420 megahertz), or a wavelength of about 21 centimeters, and with a lifetime of the energy level that undergoes the transition of 10 million years. In fact, this "21-centimeter line" of hydrogen has been of great importance in astronomy, allowing radio astronomers to chart the structure and velocity of vast clouds of atomic hydrogen in our galaxy and in others. The way to make a clock is to use this transition to control the frequency of some external electronic device. This control is achieved by preparing hydrogen atoms in the excited, or upper energy level, and injecting them into an evacuated bulb (typical dimension: about 6 inches in diameter), where they can be stored without being forced to decay to the lower level by collisions with the walls of the bulb or with other atoms for as long as a second. This helps to minimize the spread in frequencies emitted when they do decay. The atoms can then be induced to decay by "tickling" them with a radio signal from the external electronics, provided the signal is tuned precisely to the required frequency, within the narrow allowed spread. In this way, the external electronics can always be locked into, or "slaved against" the 1,420 megahertz frequency. The spread in frequencies is so narrow that the actual frequency is known to twelve significant digits, or to an accuracy of 1 part in 100 billion. Furthermore, this device maintains the same frequency to 1 part in a million billion over periods as long as days.

The original plan was to put one of these clocks in orbit, but by 1970, it was evident that the cost of the Titan 3C rocket and the 2,000-pound payload required to achieve this was more than the NASA budget would permit, and a more modest plan was developed. This was to send the clock on a suborbital flight to an altitude of about 10,000 kilometers (one and a half times the radius of the Earth) using a cheaper Scout D rocket, and a smaller payload, on the order of a few hundred pounds. There were two problems that had to be overcome to make the experiment work. The first was to build a lightweight maser clock that would withstand the 20 g's of acceleration it would experience during launch. The second problem was how to detect the gravitational red shift. Consider what happens during ascent of the rocket, say, when the rocket clock emits its signal, and the signal is received at the ground and compared with the frequency of the ground clock. The received frequency differs from the ground clock frequency because of two effects: the gravitational blue shift, caused by the height difference, and the time dilation, caused by the rocket's rapid motion. However, the received frequency is also shifted toward the red because of the usual Doppler shift produced by the rocket's motion away from the ground clock (during descent of the rocket, this effect would be a blue shift), and this Doppler shift is 100,000 times larger than the gravitational red shift for a typical Scout D velocity of several kilometers per second. We would like to eliminate this huge effect somehow, in order to see the much smaller effects of interest. This was done in a very elegant way as follows: Suppose a signal is emitted from the ground clock toward the rocket clock (this is called the "uplink," while a downward signal is called the "downlink"). When received by the rocket clock, the received frequency differs from that of the rocket clock by the Doppler shift, and by the gravitational red shift and time dilation. Incorporated into the rocket payload is a transponder, a device that takes a received signal and sends it right back with the same frequency (and with a little more power, to make up for any losses during transmission up). When the transponded signal is received back on Earth, its frequency is further red shifted by the Doppler effect, because the transponder is receding from Earth, but it is now gravitationally blue shifted by an amount that exactly cancels the gravitational red shift experienced by the signal on the uplink, and is changed by time dilation in a way that also cancels that experienced on the uplink. Therefore, when received back at the ground, this two-way signal has had its frequency changed by twice the Doppler shift, and that's all. The one-way downlink signal sent by the rocket clock and received at the ground has been changed by only one factor of the Doppler shift and by the gravitational blue shift and the time dilation. All one has to do then is take the frequency change on the two-way signal, divide by two, and subtract it from the one-way frequency change, and presto: no Doppler effect. This Doppler-cancelation scheme in fact was incorporated directly into the electronics that gathered the data from the two radio links, and so it disappeared from the experiment altogether (see figure 3.4).

After the years of development of the clocks, of making one of them space-worthy, of testing and retesting them to simulate launch conditions, the time had come to actually do the experiment. Vessot was in charge of the rocket clock at Wallops Island, while Levine took care of the ground clock at Merritt Island. As is often the case, the period leading up to launch was not without its crises. A misbehaving monitor designed to keep track of conditions in the rocket clock was brought into line by Vessot by the elegant technique of dropping it on the floor.

Finally, the countdown reached the end, and at 6:41 A.M. eastern standard time, June 18, 1976, the Scout D roared into the Virginia skies. At 6:46, the payload containing the clock separated from the fourth stage of the rocket, and was in free fall thereafter. At this point data could be taken, because the rocket clock was no longer affected by the high accelerations and vibrations of launch. For about three minutes, the one-way downlink frequency from the rocket clock (with the Doppler piece canceled automatically, remember) was lower than that of the ground clock, because the high velocity of the rocket caused a time-dilation red shift to lower frequencies, while the altitude

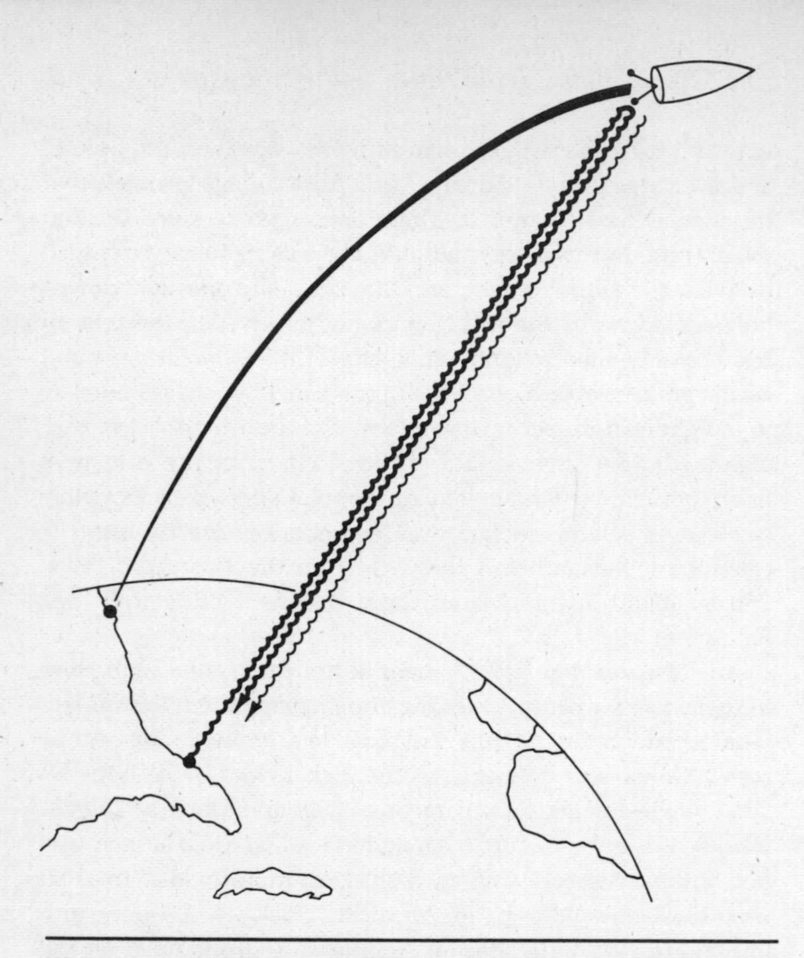

Figure 3.4 Rocket red-shift experiment. As a hydrogen maser clock sent aloft on a Scout D rocket rises over the Atlantic, a signal from an identical ground maser clock is sent toward it. When the signal is received by the rocket, it is sent back, and a signal directly from the rocket clock is sent along with it. Because the rocket clock is at a different height and is moving at a different velocity than the ground clock, the frequency of the one-way signal received by the ground clock is changed by the Doppler effect, the gravitational red shift, and the time dilation. But for the two-way signal, the ground clock is both emitter and receiver, so there is no gravitational red shift or time dilation because the signal is emitted and received at the same height and velocity. However, the Doppler shift contributes twice. The device that turns the signal around, called a transponder, sees a signal red shifted by the Doppler effect because it is moving away from the ground, and the ground receiver sees the signal Doppler shifted again because the transponder that turned it around is moving away from the ground. Thus, if one-half of the frequency change of the round-trip signal is subtracted from the frequency change of the one-way signal, the Doppler effect will cancel out.

was not yet large enough to produce a gravitational blue shift. At 6:49 the frequencies of rocket and ground clock were exactly the same: the gravitational blue shift canceled the time dilation red shift. After that, as the altitude increased and the speed of the rocket decreased, the gravitational blue shift dominated more and more. The peak of the orbit, the apogee, occurred at 7:40. Here the shift was predominantly the gravitational blue shift, amounting to almost 1 hertz out of 1,420 megahertz, or 4 parts in 10 billion. Because both the rocket clock (after separation from the fourth stage of the Scout D) and the ground clock maintained their intrinsic frequencies stably to 1 part in a million billion, they could measure these changes in frequency to very high accuracy. Data taking continued during descent, with the cancelation between gravitational blue shift and time dilation occurring again at 8:31. At 8:36, the payload was too low in the sky to be tracked reliably, and, shortly thereafter, some 900 miles east of Bermuda, this valiant little timepiece, which had done so much for our understanding of curved space-time, came to a wet and ignominious end.

This two-hour flight produced more than two years of data analysis for Vessot and his colleagues, but when all was said and done, the predicted frequency shifts (gravitational plus time dilation), based on the speed and location of the rocket known at all times through tracking, agreed with the observed shifts to a precision of 70 parts per million, or to 7/1000 of 1 percent.

In the early days of general relativity, it was felt by some physicists that the failure to confirm the gravitational red shift was a serious stumbling block to full acceptance of the theory. We now know that the gravitational red shift is not really a true test of general relativity itself, but, like the Eötvös experiment, it is a test of the principle of equivalence and of the basic notion of curved space-time. But it took the advanced technology of the 1960s and 1970s to provide the means finally to perform these experiments to high precision. Their success gives us great confidence that the idea that space-time is curved is correct. This still does not tell us that general relativity is right, however. For that we must look at experiments that test the actual predictions of the theory, predictions such as "how much" space-time curvature there is for a given gravitational field. The remainder of this book will be devoted to these tests. We begin with a test that made the name Einstein a household word.

4

The Departure of Light from the Straight and Narrow

THE HEADLINE in the London *Times* of November 7, 1919 read "Revolution in Science/ New Theory of the Universe/ Newtonian Ideas Overthrown." It heralded a brave new world in which the old values of absolute space and absolute time were lost forever. To some emerging from the devastation of the Great War, it meant the overthrow of all absolute standards, whether in morality or philosophy, music or art. In a recent survey of twentieth-century history, the British Historian Paul Johnson argued that the "modern era" began not in 1900, not in August 1914, but with the event that spawned this headline.

The event made Einstein a celebrity. Forget for a moment his genius, the triumph of his theories, and the new scientific order he created almost singlehandedly. That alone might have been enough, but Einstein was also, in today's terminology, a very "mediagenic" fellow. His absentmindedness, his playful wit, his willingness to expound upon politics, religion, and philosophy in addition to science, his violin playing—all these characteristics sparked an intense curiosity on the part of the public. The press, tired of printing battle reports from the war, was only too eager to satisfy its readers' curiosity.

The particular event that caused such a commotion was the successful measurement of the bending of starlight by the Sun. The amount of bending agreed with the prediction of Einstein's general theory of relativity, but disagreed with the prediction of Newton's gravitational theory.

The story of the deflection of light is one of the most fascinating in all science. It actually has its roots in the eighteenth century, yet the story continues to evolve to this day. It journeys from the heights of theoretical and experimental accomplishments to the depths of racist propaganda, from our solar system to the most distant galaxies.

It is believed that the first person to consider seriously the possible effect of gravity on light was a British amateur astronomer, Reverend John Michell. Ever since the time of Newton, it had been assumed that light consisted of particles or "corpuscles"; in 1783, Michell reasoned that light would be attracted by gravity in the same way that ordinary matter is attracted. He noted that light emitted outward from the surface of a body such as the Earth or the Sun would be reduced in velocity after traveling great distances (Michell, of course, did not know the theory of special relativity, which requires the speed of light to be constant). How large would a body of the same density as the Sun have to be in order that light emitted from it would be stopped by gravity and pulled back before reaching infinity? The answer he obtained was 500 times the diameter of the Sun. Light could never escape from such a body. This remarkable idea describes what we now refer to as a black hole. Fifteen years later, the great French mathematician Pierre Simon Laplace performed a similar calculation. Although Michell and Laplace were wrong in the details, their basic premise is right: Gravity affects light.

Prompted by Laplace's speculations, a Bavarian astronomer

named Johann Georg von Soldner (1776–1833) asked the related question: Would gravity bend light? This was over one hundred years before Einstein. Soldner was a largely self-taught man who became a highly respected astronomer. He made fundamental contributions to the field of precision astronomical measurements known as astrometry, and eventually rose to the position of director of the observatory of the Munich Academy of Sciences. But in 1801, he was still an assistant to the astronomer Johann Bode in the Berlin Observatory. Assuming that light was a corpuscle undergoing the same gravitational attraction as a material particle, Soldner determined how much bending would occur for a path that skimmed the surface of the Sun. According to Newtonian gravity, the orbit of one body about another is a conic section, the figure formed by the intersection of a cone with a plane tilted at various angles: an ellipse or a circle if the orbit is bound, so that the body never escapes; a hyperbola if it is unbound; or a parabola if it is borderline between the two (see figure 4.1). In this case, because the speed of light is so large, the effect of gravity is small, and the orbit is a hyperbola that is very close to being a straight line. However, the deviation, while small, is calculable, and Soldner's value was 0.875 seconds of arc (3,600 arcseconds = 1°).

This beautiful, highly original work was published in 1803 in one of the German astronomical journals. It was then immediately forgotten. It was forgotten partly because the effect was just beyond the current limits of telescope precision, and partly because of the rise during most of the nineteenth century of the wave theory of light, according to which light moves as a wave through an imponderable "aether," and therefore suffers no deflection. Einstein was certainly not aware of Soldner's paper. It was not until 1921 that Soldner's work was rediscovered and resurrected, but then it was for a different, more unsavory purpose.

Like Soldner a century before, Einstein in 1907 was interested in the effect of gravity on light. He recognized that if the principle of equivalence led to an effect on the frequency of light, the gravitational red shift, it should also result in an effect

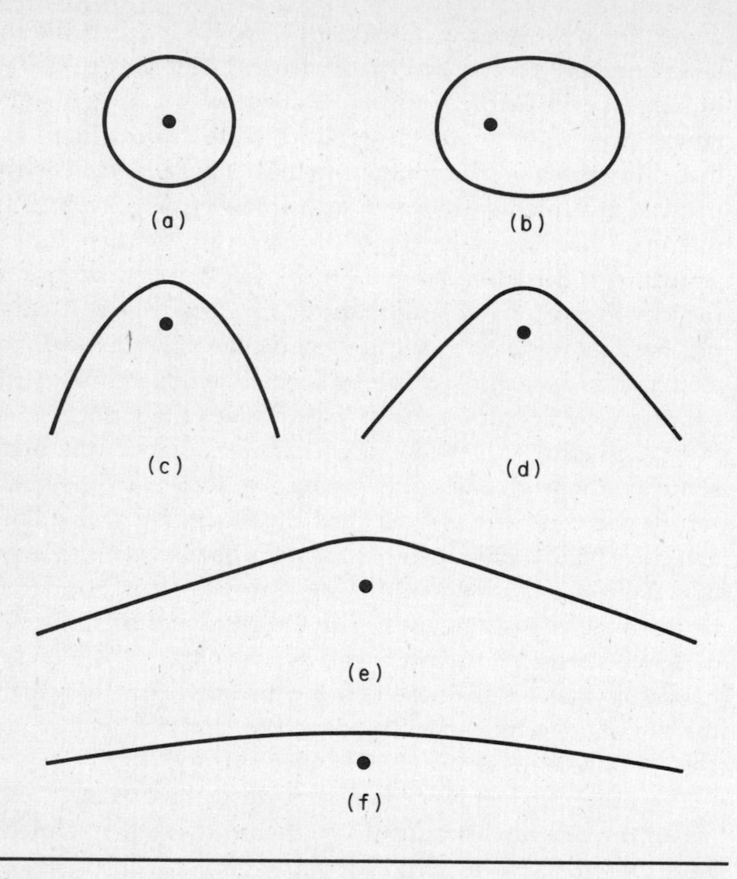

Figure 4.1 Newtonian orbits. Orbits (a) and (b) are bound orbits; orbits (c) through (f) are unbound. (a) Circular orbit. (b) Elliptical orbit. (c) Parabolic orbit. (d) Hyperbolic orbit. (e) and (f) Hyperbolic orbits of larger and larger velocity.

on its trajectory. In 1911, he determined that the deflection of a ray grazing the Sun should be 0.875 arcseconds. He proposed that the effect be looked for during a total solar eclipse, during which stars near the Sun would be visible and any bending of their rays could be detected through the displacement of the stars from their normal positions. But the effect was still near the limit of detectability, and despite the efforts of astronomer Erwin Freundlich to mount such an expedition, little interest was generated. The Freundlich expedition fell through, and Einstein's proposal was forgotten.

It is quite easy to see how Einstein's equivalence principle leads to a deflection of light, although the argument presented here is not the same as the one used by Einstein in his 1911 paper. Imagine a pair of laboratories with glass sides, each containing an observer well versed in the equivalence principle (see figure 4.2). The laboratories are perfectly constructed with parallel faces and right angles at all corners. They are connected by a frictionless trolley that keeps them parallel to each other, but allows them to move independently and freely in the parallel direction. With a blast of rockets, the laboratories take off from the Sun in a radial outward direction on their way to intercept a light signal that is about to pass by the Sun. After the rockets turn off, the laboratories are in free fall, and begin to slow down under the gravitational pull of the Sun. Just as with our previous thought experiment, their initial velocities have been cunningly chosen so that, at the moment the light ray crosses the left pane of laboratory 1, the laboratory happens to be at rest with respect to the Sun, just about to begin its backward fall. The second laboratory's velocity has also been cleverly chosen so that at the moment the light ray leaves laboratory 1 and crosses the left pane of laboratory 2, the latter is at rest with respect to the Sun and about to fall back. Now, in laboratory 1, the observer notes that the light ray enters his window at a certain location and at a certain angle. The ray then crosses the laboratory and exits through the right pane at exactly the same angle. This is completely natural according to observer 1, because he is in a freely falling frame in which gravity is apparently absent. Light must travel in a straight line in such a frame, and so must subtend equal angles on both sides of the laboratory. Observer 2 comes to a similar conclusion. She observes the ray enter her laboratory at some angle and leave at the same angle, and she too acknowledges that this is in accord

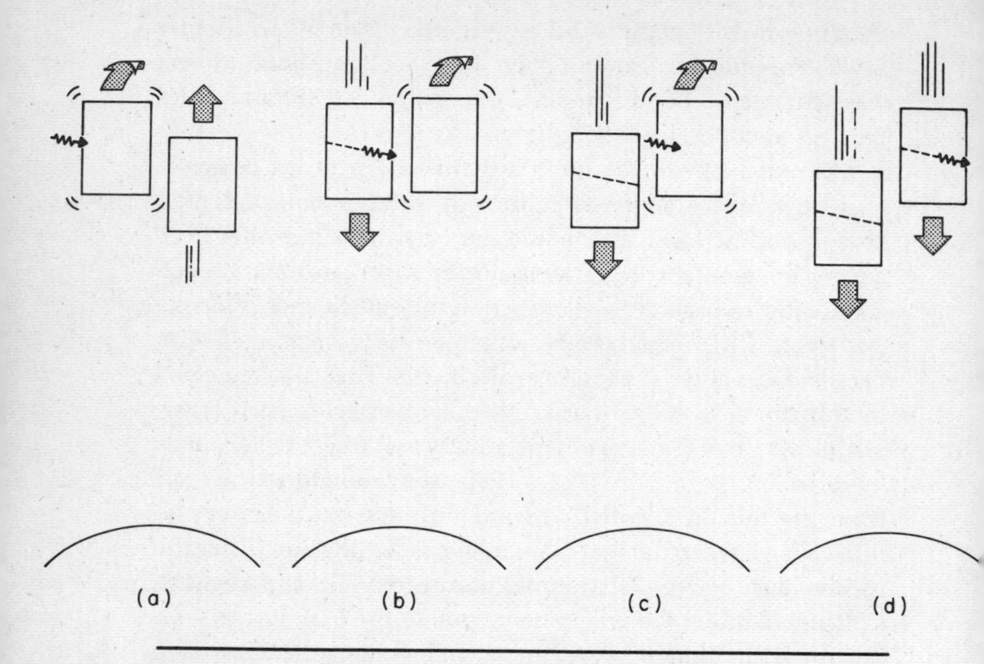

Figure 4.2 Equivalence principle and the deflection of light. A pair of laboratories is shot out from the center of the Sun, the rockets are turned off, and they are in free fall. (a) Laboratory 1 comes to rest momentarily at the top of its trajectory at the moment the light ray enters it. Laboratory 2 is still ascending. (b) Because laboratory 1 is in free fall, light moves on a straight line because of the equivalence principle, and leaves it at the same angle with which it entered. By that time, laboratory 1 has started to fall, while laboratory 2 has reached the top of its trajectory. (c) The light ray enters laboratory 2, but because laboratory 1 is falling relative to 2, the observed angle of entry is downward relative to that measured in 1, because of aberration. (d) Light moves in a straight line through laboratory 2 because it is in free fall. When these incremental deflections are added up for a sequence of laboratories along the light path, the result is 0.875 arcseconds for a grazing ray.

with the equivalence principle. However, when the observers return to base and compare their data, they find that they disagree on the angle the ray subtended in each of their laboratories. In laboratory 2, the observed angle was more in a downward direction (toward the Sun) than in 1. After a little thought, they understand why. Although each laboratory was at rest with respect to the Sun at the moment the light ray entered it, the two moments were not the same. By the time the ray crossed laboratory 1 and entered 2, laboratory 1 had begun to fall, and had picked up a velocity given by the acceleration of gravity times the time taken for light to cross it. Thus, observer 2 saw light enter her laboratory from a laboratory that was moving downward relative to her. Thus, the angle of entry was deflected downward, by the phenomenon of aberration. This is the same phenomenon as the one that causes the fronts of your legs to become soaked when you carry an umbrella quickly through an otherwise vertical rainfall. The direction in which you see the rain approach your leg is changed from the vertical because of your motion relative to the ground. Light is affected by this phenomenon as well, only to a smaller degree because of its large velocity. The upshot is that observer 2 claims that the light ray has been deflected slightly toward the Sun.

By considering a sequence of such laboratory experiments performed all along the trajectory of the light ray and adding up all the tiny deflections, the observers conclude that the net deflection of a ray that just grazes the Sun would be 0.875 arcseconds. Therefore, whether we use the Newtonian theory of gravity combined with the corpuscular theory of light, as Soldner did, or the principle of equivalence, as in this derivation, we predict the same deflection of light.

Yet in November, 1915, Einstein doubled the prediction. By that time, he had completed the full general theory of relativity, and found that, in a first approximation to the equations of the theory, the deflection had to be 1.75 arcseconds, not 0.875 arcseconds.

Was this doubling completely arbitrary? Were the previous calculations completely wrong? Not at all. They are correct as far as they go. They simply did not, indeed could not, take into account an important circumstance that only the complete general theory of relativity could cope with: the amount of curvature of space. As we have already seen, the principle of equivalence tells us that space-time must be curved, but it does not tell us in detail by how much it is curved, and in particular, it says nothing about the curvature of space itself. The curvature of space modifies the result in the following way. Imagine a large number of small, perfectly straight rulers with ends cut at perfect right angles (see figure 4.3). Take about a third of these rulers and line them up end to end, each parallel to its neighbor, stretching across a large region of empty space far from the Sun. This is the line *BO* in the figure. Now take another third of the rulers, and line them up end to end starting from point *O* again staying far from the Sun, but in such a direction that the two free ends of the lines of rulers end up on opposite sides of the Sun. Take most of the remaining rulers, and line them up end to end in the same way, starting at *B*, so that they go just past the surface of the Sun, and on to point *A* to complete a large triangle. Two sides of the triangle (*OA* and *OB*) are far from the Sun, while the third side (*AB*) grazes its surface. With the remaining rulers, start at a point, *B'*, setting off at the same angle as you set off at *B*, and complete the triangle *OA'B'*.

General relativity predicts that space is curved near gravitating bodies, the curvature being greater the closer one gets to the body, and negligible at large distances. Because the triangle *OA'B'* is far from the Sun, the sum of its interior angles is 180°, just as in ordinary flat space. But a detailed calculation using the equations of general relativity reveals that the sum of the interior angles of the triangle *OAB* is no longer 180° (remember the Spherelanders) but is 179°, 59 minutes, 59.125 seconds, or 180° less 0.875 arcseconds!

This explains Einstein's doubling. The previous calculations, such as the one using the freely falling laboratories, gave the deflection of light relative to locally straight lines; remember that the laboratories were lined up parallel to each other using the same kinds of local straight rulers and right angles as were the rulers in our big triangles. If we thought that space was flat, that would be it. However, general relativity predicts that locally straight lines that pass near the Sun are bent relative to straight lines that pass far from the Sun in completely empty

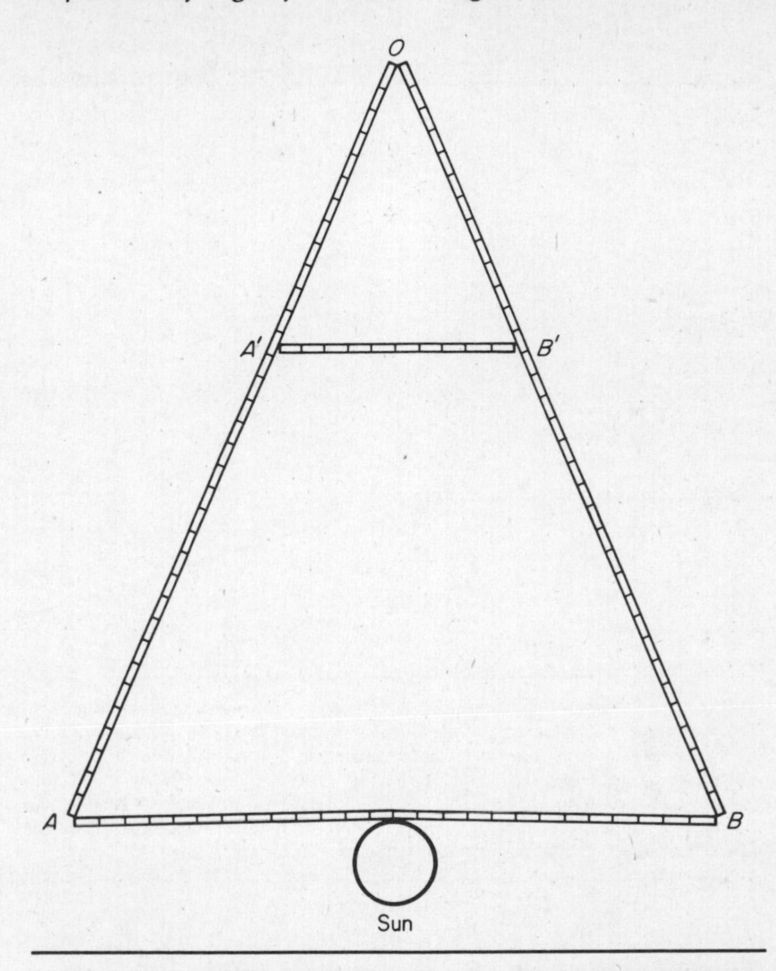

Figure 4.3 Solar-system triangle. The line of rulers *BO* and *OA* are far from the Sun, while the line of rulers *AB* grazes the Sun. The line of rulers *A′B′* is far from the Sun, and the angle *OB′A′* is the same as the angle *OBA*. The sum of the interior angles of *OA′B′* is 180°. The sum of the interior angles of *OAB* is 179°, 59 minutes, 59.125 seconds. Thus, angle *OAB* is 0.875 arcseconds smaller than angle *OA′B′*.

space by an additional 0.875 arcseconds (see figure 4.4). Thus, the total deflection must be 1.75 arcseconds.

As we shall see later, this curvature effect is the important

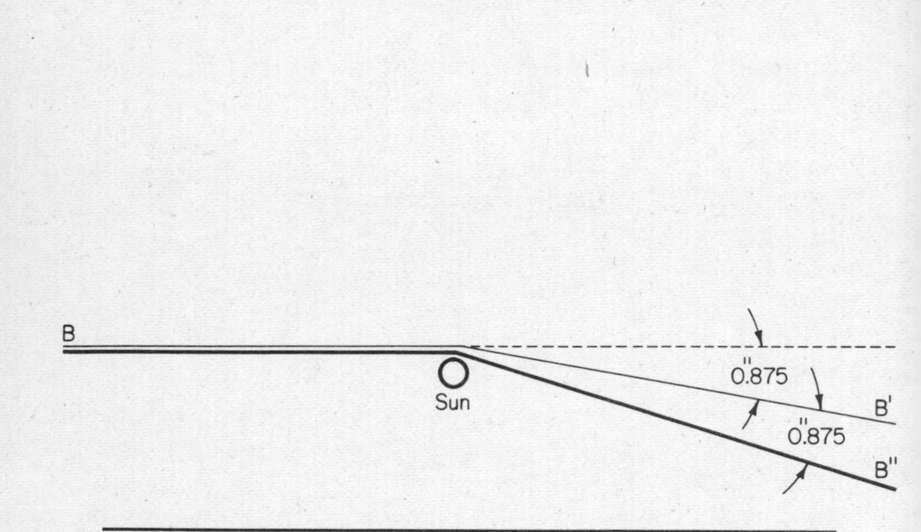

Figure 4.4 Combined light deflection. Line AA': A straight line far away from the Sun. Line BB': A "straight" line of rulers grazing the Sun, deflected relative to distant line by 0.875 arcseconds (according to general relativity). Line BB": Direction of light ray, deflected relative to "straight" line BB' by 0.875 arcseconds. This part comes from the principle of equivalence or from Newtonian theory. The deflection of grazing rulers relative to distant rulers depends on the amount of space curvature, and can vary from theory to theory.

difference in the predictions of different theories of gravity. Any theory of gravity that is compatible with the equivalence principle predicts the first 0.875 arcseconds part. The second part comes from space curvature. Newtonian theory is a flat-space theory, so there is no further effect; the prediction remains at 0.875 arcseconds. General relativity, purely by coincidence, predicts an amount of curvature that just doubles the deflection. The alternative Brans-Dicke theory (described in more detail in chapter 8) predicts slightly less curvature than general relativity, resulting in a slightly smaller value for the second part and a slightly smaller total deflection.

Einstein's doubling of the predicted deflection had important consequences, for it meant that the effect was now accessible to observation. But the fact that a successful observation came as early as 1919, only three years after publication of the general theory, must be credited to the pivotal role played by Sir Arthur Stanley Eddington (1882–1944). By the time of the outbreak of World War I, Eddington held a position of prominence in British science as one of the foremost observational astronomers of the day, and had recently been elected a Fellow of the Royal Society and appointed the Plumian Professor at Cambridge University. The war had effectively cut off direct communication between British and German scientists, but the Dutch cosmologist Willem de Sitter managed to forward to Eddington Einstein's latest paper together with several of his own on the general theory of relativity. Eddington recognized the deep implications of this new theory, and he immediately set out to learn the mathematics required to master it. In 1917, he prepared a detailed report on the theory for the Physical Society of London. This helped to spread the word. Eddington and Astronomer Royal Sir Frank Dyson also began to contemplate an eclipse expedition to measure the predicted deflection of light. As an astronomer at the Royal Greenwich Observatory from 1906 to 1913, Eddington had made an eclipse expedition in 1912 and was familiar with the techniques and problems involved. Dyson had pointed out that the eclipse of May 29, 1919 would be an excellent opportunity because of the large number of bright stars expected to form the field around the Sun. A grant of 1,000 pounds sterling was obtained from the government, and planning began in earnest. The outcome of the war was still in doubt at this time, and a danger arose that Eddington would be drafted. As a devout Quaker, he had pleaded exemption from military service as a conscientious objector, but, in its desperate need for more manpower, the Ministry of National Service appealed the exemption. Finally, after three hearings and a last-minute appeal from Dyson attesting to Eddington's importance to the eclipse expedition, the exemption from service was upheld on July 11, 1918. This was just one week before the second Battle of the Marne.

In our present time, when cold-war politics sometimes obstructs the free flow of scientific information and interaction, we would do well to remember this example: a British government permitting a pacifist scientist to avoid wartime military duty so that he could go off and try to verify a theory produced by an enemy scientist.

On March 8, 1919, just four months after the end of hostilities, two expeditions set sail from England: Eddington's for the island of Principe, off the coast of Spanish Guinea; the other team under Andrew Crommelin for the city of Sobral, in northern Brazil. The principle of the experiment is deceptively simple. During a total solar eclipse, the moon hides the Sun completely, revealing the field of stars around it. Using a telescope and photographic plates, the astronomers take pictures of the obscured Sun and the surrounding star field. These pictures are then compared with pictures of the same star field taken when the Sun is not present. The comparison pictures are taken at night, several months earlier or later than the eclipse, when the Sun is nowhere near that part of the sky and the stars are in their true, undeflected positions. In the eclipse pictures, the stars whose light is deflected would appear to be displaced away from the Sun relative to their actual positions. One property of the predicted deflection is important: Although a star whose image is at the edge of the Sun is deflected by 1.75 arcseconds, a star whose image is twice as far from the center of the Sun is deflected by half as much, and a star 10 times as far is deflected by one-tenth; in other words, the deflection varies inversely as the angular distance of the star from the Sun. Now, because the eclipse pictures and the comparison pictures are taken at different times, under different conditions (and sometimes using different telescopes), their overall magnifications may not be the same. Therefore, the stars on the plates that are farthest from the Sun, undeflected on the comparison plate, deflected only negligibly on the eclipse plate, can be used to determine an overall magnification correction. Then the true deflection of the stars closest to the Sun can be measured.

In practice, of course, nothing is ever this simple. One important complication is a phenomenon the astronomers call "seeing." Because of turbulence in the earth's atmosphere, starlight passing through it can be refracted or bent by the warmer and colder pockets of moving air and can suffer deflections of as much as a few arcseconds (this is part of what makes stars twinkle to the naked eye). These deflections are comparable to the effect being measured. But because they are random in nature (as likely to be toward the Sun as away from it), they can be averaged away if one has many images. The larger the number of star images, the more accurately this effect can be removed. Therefore, it is absolutely crucial to obtain as many photographs with as many star images as possible. To this end, of course, it helps to have a clear sky.

We can therefore imagine Eddington's emotional state when, on the day of the eclipse, "a tremendous rainstorm came on." As the morning wore on, he began to lose all hope. Before the expedition, Dyson had joked about the possible outcomes: No deflection would show that light was not affected by gravity, a half-deflection would confirm Newton, and a full-deflection would confirm Einstein. Eddington's companion on the expedition had asked what would happen if they found double the deflection. Dyson had answered, "then Eddington will go mad, and you will have to come home alone." Now Eddington had to consider the possibility of getting no results at all. But at the last moment, the weather began to change for the better: "The rain stopped about noon, and about 1:30, when the partial phase [of the eclipse] was well advanced, we began to get a glimpse of the Sun." Of the sixteen photographs taken through the remaining cloud cover, only two had reliable images, totaling only about five stars. Nevertheless, comparison of the two eclipse plates with a comparison plate taken at the Oxford University telescope before the expedition yielded results in agreement with general relativity, corresponding to a deflection for a grazing ray of 1.60 ± 0.31 arcseconds, or 0.91 ± 0.18 times the Einsteinian prediction. The Sobral expedition, blessed with better weather, managed to obtain eight usable plates showing at least seven stars each. The nineteen plates taken on a second telescope turned out to be worthless because the telescope apparently changed its focal length just before totality of the eclipse, possibly as a result of heating by the Sun. Analysis of the good plates yielded a grazing deflection of 1.98 ± 0.12 arcseconds, or 1.13 ± 0.07 times the Einsteinian value.

Despite the difficulties and disappointments, these expeditions were triumphs for observational astronomy and produced a victory for general relativity. This did not imply unconditional acceptance of the theory, of course. Legitimate scientific doubts remained: compared to other physical theories of the day, general relativity was mathematically very complex, and its validity rested on only two tests to date—the deflection of light and the perihelion advance of Mercury (to be described in chapter 5). Any new theory of nature must stand the test of many theoretical and experimental checks. But there were other doubts about general relativity, nonscientific doubts of a scurrilous nature that were expressed by individuals who called themselves scientists. These doubts about the theory were based purely on the fact that Einstein was Jewish. It was here that the fine work of Soldner was resurrected, and by association, debased.

The rise of anti-Semitism in Germany between the world wars had its counterpart in scientific circles. One of the leading exponents of this view was Philipp Lenard, a Nobel Laureate in Physics (1905) for his work on cathode rays. An avowed Nazi, Lenard spent much of his time between the wars attempting to cleanse German science of the Jewish taint. Relativity represented the epitome of "Jewish science," and much effort was expended by Lenard, Johannes Stark, also a Nobel Laureate in Physics (1919), and others in attempts to discredit it. The vast majority of non-Jewish German physicists did not share this view, however, and despite the Nazi takeover in Germany and the subsequent dismissal and emigration of many Jewish physicists (including Einstein), the anti-relativity program became little more than a footnote in the history of science.

Nevertheless, early in 1921, while preparing an article against special relativity, Lenard learned of the existence of Soldner's 1803 paper. This discovery delighted him, because it showed the precedence of Soldner's "Aryan" work over Einstein's "Jewish" theory. The fact that the eclipse results favored the latter over the former did not appear to faze him. Lenard prepared a lengthy introductory essay, incorporated the first two pages of Soldner's paper verbatim and summarized the rest, and had the whole thing published under Soldner's name in the September 27, 1921 issue of the journal *Annalen der Physik.* Fortunately, most physicists were not swayed by these kinds of considerations, and instead focused their attention on experimentation, not race, as the judge of scientific theory.

Although the eclipse results distinguished clearly among the possibilities of no deflection, the Newtonian deflection, and the Einsteinian deflection, their relatively large experimental errors made it important to repeat the measurements. During subsequent eclipses, numerous expeditions attempted to do better—there were three expeditions in 1922, one in 1929, two in 1936, one each in 1947 and 1952, and most recently, one in 1973. Surprisingly, there was very little improvement, with different measurements giving values anywhere between three-quarters and one and one-half times the general relativistic prediction. The 1973 expedition is a case in point. Organized by the University of Texas and Princeton University, the observation took place in June at Chinguetti Oasis in Mauritania (eclipses have the nasty habit of occurring in out-of-the-way places). The observers had the benefit of 1970s technology: Kodak photographic emulsions, temperature control inside the telescope shed (the outside temperature at mid-eclipse was 97° F), sophisticated motor drives to control the direction of the telescope accurately, and computerized analysis of the plates. Unfortunately, they couldn't control the weather any better than Eddington could. Eclipse morning brought high winds, drifting sand, and dust too thick to see the Sun. But as totality of the eclipse approached, the winds died down, the dust began to settle, and the astronomers took a sequence of photographs during what they have described as the shortest six minutes of their lives. They had hoped to gather over 1,000 star images, but the dust cut the visibility to less than 20 percent and only a disappointing 150 were obtained. After a follow-up expedition to the site in November to take comparison plates, the photographs were analyzed using a special automated device called the GALAXY Measuring Engine at the Royal Greenwich Observatory near London. The result was a deflection of 0.95 ± 0.11 times the Einsteinian prediction, still only a modest improvement over previous eclipse measurements.

The 1973 expedition was really the swan song for this type of measurement, because much more accurate determinations of the deflection were already being made using a technique that was a marriage of two of the most important astronomical discoveries of the twentieth century: the radio telescope and the quasar.

Radio astronomy began in 1931, when Karl Jansky of the Bell Telephone Laboratories in New Jersey found that the noise in the radio antenna he was trying to improve for use in radio telecommunications was coming from the direction of the center of our galaxy. The development of radar during World War II led to new receivers and techniques and to the rapid development of radio telescopes as new astronomical tools. Among the sources of radio waves that were discovered were the Sun itself, gas clouds such as the Crab Nebula, clouds of hydrogen atoms and of complex molecules, and radio galaxies. Radio waves are the same as ordinary visible light, only of longer wavelength. Whereas visible light spans the wavelength range 4,000 to 7,000 angstroms, radio waves span the range from a tenth of a millimeter to several meters. We would expect radio waves to be deflected by a gravitating body such as the Sun in exactly the same manner as visible starlight, and indeed, general relativity predicts the same deflection, independent of wavelength.

To measure the deflection of radio waves, we need to be able to measure to high precision the direction from which they come. To this end, the radio interferometer is the ideal instrument. A radio interferometer measures the direction from which a radio wave is incident in much the same way as a blindfolded person can sense the direction of a sound. For example, a sound wave approaching a person from the right will enter the right ear first, and then a short time later will enter the left ear. The difference in time is small, only one-third of a millisecond (1,000 milliseconds = 1 second) for a sound approaching from the direction of "one o'clock," but because of it, the two sound waves will be slightly out of phase upon entering each ear. The waves can then augment each other or cancel each other out, depending on whether they are in or out of phase. This is the phenomenon of constructive or destructive interference. If the waves are only slightly out of phase, the interference will be only partial. It turns out that the ear and brain are most sensitive to this effect for the lower sound tones, below about 1,500 cycles per second, for which the wavelength of the sound is longer than the distance between the ears. For higher tones, the direction of the sound is sensed primarily by the difference in intensity detected by the two ears, at least for nearby sources.

In a radio interferometer, the role of the ears is played by two radio telescopes, and the incoming sound is replaced by an incoming train of radio waves. The difference in the time of arrival of a given wave front at the two telescopes determines the amount and pattern of interference of the radio waves when the signals from the two telescopes are combined. The time difference in turn depends on the angle at which the wave front approaches relative to a line drawn between the two telescopes, called the baseline. For a given wavelength of the radio waves, the longer the baseline, the larger the time delay for a given angle of approach, and thus the more accurately the angle can be measured (for different wavelengths, the accuracy with which the interference pattern can be interpreted changes; all other things being equal, the accuracy increases as the wavelength decreases). Unlike the human ear, the radio interferometer can operate effectively at wavelengths much shorter than the baseline. Also unlike the ear, the intensity received by the two telescopes is the same, so this effect is not important. Radio interferometers range in baseline from the 1-kilometer instrument in Owens Valley, California, to the 35-kilometer device at the National Radio Astronomy Observatory (NRAO) in West Virginia, to a 3,900-kilometer arrangement, known as "Goldstack," that combined the Goldstone, California, telescope with the Haystack telescope in Westford, Massachusetts. Intercontinental interferometers tying together as many as eighteen telescopes around the world have also been used. The resolution of some of these interferometers can approach one ten-thousandth of an arcsecond.

We also need a very sharp source of radio waves. Most astronomical sources are unsuitable for this purpose because they are extended in space. For example, most radio galaxies emit their waves from an extended region that can be as large as a degree. The discovery of quasars, besides motivating applications of general relativity to astrophysics, provided the ideal source of radio waves to test the deflection of light. Because they are so distant, on the order of 10 billion light years away, they appear much smaller in extent, making it possible to pinpoint their location more accurately. Yet despite their distance, many of them are powerful radio sources. However, a powerful point source of radio waves is not the only ingredient for a successful light deflection experiment. We need at least two of them fairly close to each other on the sky, and they have to pass near the Sun as seen from Earth. We need at least two for the same reason as we needed a field of stars behind the eclipsed Sun in the optical deflection measurements: the stars whose images are far from the Sun are used to establish the scale because their light is relatively undeflected, and the movement of the star images close to the Sun is used to determine the deflection.

Now comes a heavenly coincidence. Two of the brightest

quasars on the sky are 3C273 and 3C279 and both have a small angular size. Each October 8, they pass very close to the Sun as seen from Earth; in fact, 3C279 actually passes behind the Sun, while 3C273 comes as close as 4° (see figure 4.5). The power of radio interferometry is that it is most adept at measuring the angle between two such radio sources, with even more accuracy than it can measure their absolute positions on the sky. Over a period of ten days centered on October 8, the pair of quasars passes from one side of the Sun to the other, and each day the angle between the two can be measured on the radio interferometer. What would be the expected output of the instrument? Initially, when both quasars are far from the Sun, the angle measured would be the true, undeflected angle between them. As days pass, the pair approaches the Sun, and by October 7, 3C279 is much closer than 3C273. Because the amount of deflection varies inversely as the distance of the light path from the Sun, the radio waves from 3C279 are deflected more than those from 3C273, and the angle between the two will appear to increase. As 3C279 just grazes the Sun, its light is deflected by about 1.75 arcseconds, while 3C273, 9° away and therefore 35 times farther from the Sun, is deflected only 0.05 arcseconds. By October 9, 3C279 is on the same side of the Sun as 3C273, so the angle between the two appears to decrease. As days continue to pass, the pair moves farther from the Sun and their separation returns to the original value.

By careful analysis of the behavior of the separation angle with time, the radio astronomers can determine the deflection of light as a function of distance from the Sun and translate that into a deflection of a grazing ray. One of the advantages of this technique is that it can be done every year, as opposed to eclipse measurements, which must occur sporadically and in inhospitable locales. Each October, as in some ancient harvest ritual, one can imagine the radio astronomers marching out to the interferometer to measure the bending of light. The first successful ritual occurred in 1969; in fact, there were two, one at Owens Valley, and the other at Goldstone. The results were in

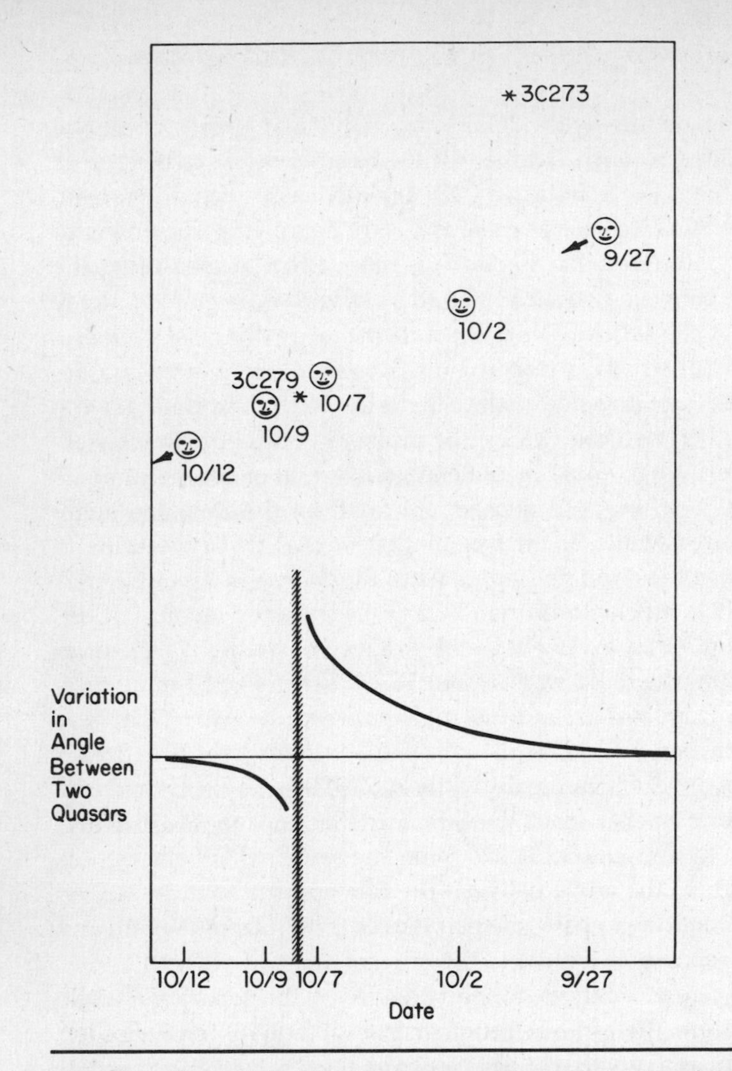

Figure 4.5 Quasar light deflection measurements. Upper portion shows the quasars 3C273 and 3C279 and the apparent path of the Sun between September 27 and October 12 each year. The Sun actually passes in front of 3C279. Lower portion shows the variation in the apparent angle between them that would be measured by a radio interferometer. September 27: very little change from the normal undeflected angle because the Sun is not close to either quasar. October 2: small increase in angle because both apparent quasar positions are deflected slightly away from the Sun. October 7: large increase in angle because the 3C279 position is deflected a large amount away from the Sun. October 8: no data because 3C279 is behind the Sun. October 9: modest decrease in angle because the 3C279 position is deflected away from the Sun, that is, toward 3C273. October 12: return to nominal undeflected angle.

agreement with general relativity, with experimental uncertainties of between 10 and 15 percent, not much better than those from eclipse expeditions. But this was only a first try. Each year the measurement could be repeated, techniques could be improved, sources of error eliminated or controlled, and more sophisticated data analysis used. This is exactly what happened: every year between 1969 and 1975, at least one measurement of the deflection of quasar radio waves was carried out, at telescopes ranging from Goldstone to Haystack to NRAO to Mullard in England to Westerbork in the Netherlands. The accuracies steadily improved, to 8 percent, to 5 percent, to 3 percent, to 1 percent, with all but one measurement in agreement with general relativity. One of the 1970 measurements was about 10 percent low, but because it was never confirmed by any other observation, it has been given little credence. The final two observations, in 1974 and 1975, did not use 3C273 and 3C279. They used a trio of quasars, called 0111+02, 0116+08, and 0119+11, that lie on almost a perfect straight line in the sky and pass by the Sun in April. The middle member of the trio, 0116+08, passes behind the Sun.

The 1975 measurement was the last attempt to measure directly the deflection of radio waves by the Sun for several years. By that time the results had reached the limits of accuracy inherent in the method, and most of the radio astronomers involved were not eager to make further major efforts to better the accuracy. The primary factor limiting the accuracy was the solar corona, the hot, turbulent gas of ionized hydrogen at 2 million degrees that extends out to several solar radii from the Sun. When radio waves pass through this gas, they are deflected from their path by refraction, the same phenomenon that causes light to change course on passing from one medium to another, such as from water to air, making the fish you are trying to spear look higher than it is. The size of the coronal deflection is larger at longer radio wavelengths, and smaller at shorter wavelengths. For a wavelength of 3.7 centimeters, a typical value for quasar observations, the deflection of a ray passing at three solar radii due to the corona is about one-quarter of the Einstein prediction, a sizable effect, and for rays that pass closer to the Sun, it is even worse. (At the shorter wavelengths of optical light, this effect is negligible, so it was not a problem for the eclipse measurements.) It is possible, by devising models for the corona, by measuring the deflection of radio waves at several different wavelengths, or simply by not looking at the data for rays that pass through the densest part of the corona, to control this effect down to the 1 percent level. One recent example of this was a series of transcontinental and intercontinental interferometric quasar observations made primarily to monitor the Earth's rotation; a test of relativity was a by-product. Because the accuracy of the interferometers was two ten-thousandths of an arcsecond, they could detect the deflection even when the quasars were 90° from the Sun, where it amounts to four-thousandths of an arcsecond, in fact all the way to 175° from the Sun. In other words, at current levels of precision of radio interferometry, the deflection of light is significant virtually everywhere in the sky. Therefore, observations close to the Sun and its corona were not crucial, and the result was a 1 percent test of the bending. Nevertheless, there are enough uncertainties left, particularly having to do with turbulent fluctuations in the density of the coronal gas, that it has proved difficult to go significantly below the 1 percent level. Still, this is a significant improvement over the 10 to 20 percent accuracies achieved by the optical measurements.

In 1979, the story of the deflection of light took a new and interesting turn with the discovery of the "double quasar." This system, listed in astronomical catalogues as Q0957+561 was a pair of quasars separated in the sky by about 6 arcseconds. This by itself would not have been so unusual were it not for the fact that the two quasars were uncannily similar: their recession velocities were identical, within the precision of the measurement, and their spectra were almost identical. The only apparent difference was that one member of the pair was somewhat fainter than the other. The astronomers who discovered this system using telescopes of the University of Arizona and Kitt Peak National Observatory immediately proposed an explanation. They argued that there was actually only one quasar and that somewhere along the line of sight between us and it was a massive object that was deflecting the light from the quasar in such a way as to produce the multiple images. The subsequent detection of a faint galaxy between the two quasar images along with a surrounding cluster of galaxies confirmed this interpretation. At least six other such multiple quasars have been found since then.

The idea that a massive object could produce an image by "gravitational lensing" was not new. The physicist Sir Oliver Lodge first suggested it shortly after the eclipse confirmations of the deflection of starlight in 1919. In the 1930s Einstein himself, as well as the astronomer Fritz Zwicky, considered the question in detail. In the 1960s, a flurry of theoretical activity on gravitational lenses took place. The actual discovery of gravitational lenses has given general relativity a new role in astronomy. The idea is not to use the gravitational lens as a new test of general relativity. Instead, the idea is to treat the general relativistic description of the deflection of light as correct (based, of course, on the previous observations that I have described), and to use the lens effect as an astronomical tool. For example, the number of quasar images, their relative brightness and placement, and any distortion in their shape (especially of the extended radio lobes that accompany most quasars) all depend in detail on the distribution of matter in the intervening galaxy or cluster of galaxies. Thus, we can learn something about the structure of galaxies by this method. Also, because the different images are produced by light rays that have traveled along different paths, an outburst of light in the quasar will be seen at different times in the different images. Compared to the billions of years it takes for the light from the quasar to reach us, the predicted differences—on the order of years—are very small, but by our standards, they are observable given enough patience and diligence. The precise time difference depends on the distribution of matter in the lensing galaxy and on the distance to the quasar. This method could give us new information about the distances to these unusual objects.

Ilse Rosenthal-Schneider, one of Einstein's students in 1919, was amazed at his remarkably serene reaction to the telegram from Eddington announcing the eclipse results. When she asked how he would have felt if the observations had not confirmed his prediction, he answered: "Then I would have been sorry for the dear Lord—the theory is correct." Einstein, of course, was joking. He understood full well that a theory stands or falls on the basis of its agreement with experimentation. Yet to his mind, general relativity was so beautiful, so elegant, so internally consistent that it had to be correct. The eclipse results merely justified his already supreme confidence. Some of that confidence was also based on another prediction of the theory that he had made even before calculating the full light bending. This prediction accounted precisely for a discrepancy in the orbit of Mercury known as the "anomalous perihelion advance" that was one of the major unsolved problems of nineteenth-century mechanics. The "perihelion of Mercury" test became one of the great underpinnings of the theory. But if Einstein could have foreseen what would happen to this test in the middle 1960s, his confidence might well have been shaken.

5

The Perihelion Shift of Mercury: Triumph or Trouble?

NEWTON'S theory of gravitation was one of the most successful physical theories of all time. Not only could it explain the gross features of the motions of the planets, such as the relation between their orbital periods and their distances from the Sun, but it could also account for the details. On the broad level, it led to a complete understanding of the laws of planetary motion, which had been derived by Johannes Kepler (1571–1630) as a way to summarize the wealth of observational data on the planets that had been accumulated and codified during the sixteenth century by Tycho Brahe and others. These laws were: (1) The planets orbit the Sun in ellipses, with the Sun at one focus; (2) the line joining the Sun and a planet sweeps through equal areas in equal times; and (3) the square of the period of the orbit of a planet is proportional to the cube of its semimajor axis (half the longest dimension of the ellipse). Newtonian gravitation theory held that the gravitational force is directed from one body to the other and varies as the inverse square of the separation of the bodies, and this together with Newton's laws of motion, was sufficient to derive Kepler's laws for any system of two bodies in orbit. One of the first successes of Newton's theory of gravitation was its use by Sir Edmund Halley to determine that the orbits of the comets that had appeared in 1531, 1607, and 1682 were actually the same orbit, and that the comet should reappear approximately every seventy-five years. The reappearance on schedule of Halley's comet in 1986 was one of the scientific media events of the year.

In reality, of course, the detailed motions of the planets were more complicated than the simple Keplerian picture, but even these complications could be explained as being due to the perturbations of a given planetary orbit by the gravitational forces due to the other planets. This picture was so successful that by the middle of the nineteenth century, John Couch Adams (1819–92) in England and independently Joseph Le Verrier (1811–77) in France were able to predict the approximate location of a hitherto unknown planet that had to be present in order to explain certain perturbations in the orbit of Uranus. Astronomical searches of the predicted location in 1846 indeed revealed a planet, which was named Neptune.

Another great achievement of Newtonian theory was the problem of the motion of the Moon. Again, in its gross features, the Moon's orbit about the Earth is an ellipse satisfying Kepler's three laws; however, its motion is strongly perturbed by the Sun. As early as 1765, a complete and accurate accounting of these perturbations could be given, with one exception. That exception was a phenomenon called the "secular acceleration," which manifests itself in a systematic increase in the mean distance of the Moon from the Earth. However, this phenomenon was, and still is, attributed to the complicated and still insufficiently understood effect of friction on the Earth's surface caused by the tides generated by the Moon, rather than to a problem with Newtonian theory itself. (We will encounter the secular acceleration of the Moon again in chapter 9 when we ask whether gravity is getting weaker.)

Thus, by 1850, the science of celestial mechanics—the study of the motions of heavenly bodies—was in great shape. It could explain every motion in the solar system that it was expected to explain, and could do so with great precision.

But a cloud soon appeared on the horizon. Following his success with Neptune, Le Verrier had been appointed director of the Paris observatory in 1854, and in 1859 he published a sophisticated theory of the motion of Mercury that revealed a major problem. Le Verrier had actually discovered this problem in 1843, but partly because his calculations were not yet as accurate as he would have liked, and partly because he had not yet achieved a position of security and prominence in the astronomical community, he had not pressed the case. But now he was more confident of his stature and his results.

The problem was this: In the absence of perturbations, the orbit of Mercury would be an ellipse whose orientation is fixed in space. One way of expressing this is to state that the perihelion—the point in the orbit that is the closest to the Sun—is fixed, so that a line drawn from the Sun to the perihelion points in a fixed direction. One of the many effects of perturbations due to the gravitational attractions of the other planets is to cause the ellipse to rotate in its plane, so that the actual orbit of the planet describes a kind of rosette pattern (see figure 5.1). Equivalently, the perihelion direction is no longer fixed, but rotates, or "precesses." This precession could be seen in the observations of Mercury's motion, and amounted to 574 arcseconds per century. In 2,250 centuries, Mercury's orbit would trace out a complete rosette. Le Verrier attempted to account for this precession using standard Newtonian celestial mechanics. The largest contribution comes from Venus, because it occupies the next closest orbit to Mercury: 277 arcseconds. Next comes the effect of Jupiter, at 153 arcseconds. Even though Jupiter is 10 times farther away from Mercury than Venus is (remember that the force it exerts on Mercury falls off as the inverse square of the distance), it is 400 times more massive than Venus. The

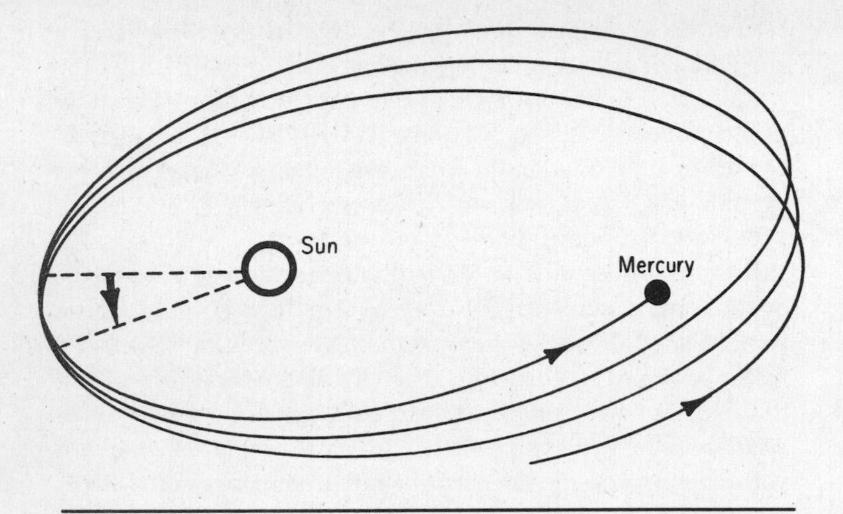

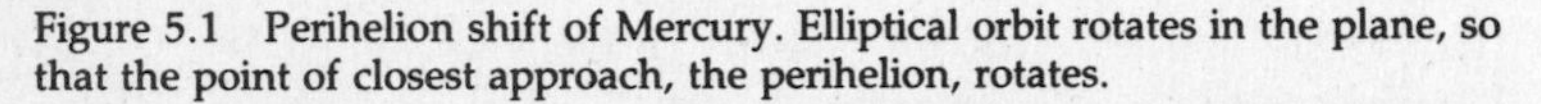

Figure 5.1 Perihelion shift of Mercury. Elliptical orbit rotates in the plane, so that the point of closest approach, the perihelion, rotates.

Earth's effect is next, at about 90 arcseconds, and Mars and the rest of the planets together contribute about 10 arcseconds. Unfortunately, the total of these contributions is only 531 arcseconds per century, 43 arcseconds too small. Actually, the value for the discrepancy obtained by Le Verrier was 38 arcseconds, but for the purposes of this discussion we will use the modern values.

This produced a real crisis in nineteenth-century science, one of many to occur during the latter half of the century that would signal the need for some new physics. Numerous proposals were put forward to account for the discrepancy. One was the existence of new matter in an orbit between Mercury and the Sun. This matter could be in the form of a planet, which had already been given the name Vulcan after the Roman god of fire, because, being so close to the Sun, it would be very hot. Or it could be in the form of a ring of dust or asteroids similar to the rings of Saturn. Given enough mass, such matter could generate the required gravitational perturbations to produce the additional perihelion precession of Mercury. Unfortunately, no solid observational evidence for Vulcan or for such a ring was ever found, despite numerous telescopic searches and several unconfirmed sightings. Another proposal was to modify the Newtonian inverse square law. One proponent of this idea was the U.S. astronomer Simon Newcomb (1835–1909), whose careful analysis of Mercury's motion first established what is the correct value of the anomalous perihelion precession, 43 arcseconds per century. Newcomb argued that the excess could be accounted for if the force law of gravitation were modified from the inverse square to the inverse power 2.0000001574. But by the turn of the century, this proposal was also dead, ruled out by a more accurate calculation of the lunar orbit, whose agreement with observation was much worse using Newcomb's modified law than it was using the pure inverse square law. From 1859 well into the twentieth century a host of suggestions was made to explain Mercury's perihelion precession, some serious, some outrageous, some simple, some very complicated, and none very successful.

In November, 1915, while struggling to put the finishing touches on the general theory, Einstein was well aware of the problem of Mercury, and it was one of the first calculations he carried out using the new theory. To his delight, he found a precession of 43 arcseconds per century! He later wrote, "for a few days, I was beside myself with joyous excitement," and told a colleague that the discovery had given him palpitations of the heart. The fact that the theory agreed so closely with the observation without any special assumptions or complicated adjustments gave him extra pleasure.

From the point of view of curved space-time, the gravitational fields in the solar system are very weak, in the sense that the deviations they produce from flat space-time are tiny. This means that the equations of general relativity can be applied to the solar system by a method of successive approximations. In the first approximation, the equations are the same as standard Newtonian theory; therefore, all the usual results for celestial mechanics are reproduced, including the perturbations due to the other planets. But in the next approximation, there are small corrections to the Newtonian formulas, called post-Newtonian corrections, meaning after or beyond the Newtonian. Loosely speaking, when applied to the motion of a planet such as Mercury, these corrections are of three types. First are curvature effects: because the space through which the planet moves is curved, distances and angles are slightly different than one would expect in ordinary flat space (this is similar to the effect of curvature on the trajectory of light rays in the deflection of light). Second are velocity effects: the effective increase in the inertial mass of a moving body predicted by special relativity (see the appendix) has its counterpart here in a velocity-dependent modification of the gravitational force. Third are nonlinearity effects: in Newtonian theory, the gravitational force is determined from a potential, while in general relativity, the force is determined from this potential minus a small term proportional to the square of the potential, the net force thus being slightly weaker than you would expect. This description of the various contributions is very loose, and depends on which representation of the mathematical equations one is using. The total effect of these post-Newtonian corrections on Mercury's orbit is an unambiguous precession of 43 arcseconds per century.

The other proposed explanations for Mercury's perihelion precession that were circulating continued to do so for a time, but gradually fell out of favor in the face of the beauty and simplicity of the Einsteinian prediction. The perihelion precession became one of the great experimental pillars of general relativity, and would remain so for half a century.

But beginning in the 1960s, two developments occurred that made the perihelion precession an even greater success for general relativity, while simultaneously casting doubt on it. The first of these was the qualitative change in the manner of observing the motions of the planets brought about by radar and by space exploration. Sending radar signals to Mercury and

Venus and measuring the time elapsed before receiving the echo (radar ranging) gave significant improvements in the accuracy of determinations of their orbits, and gave more accurate values for the total perihelion precession of Mercury. Encounters of various spacecraft with Venus and Mars and analysis of their orbits by and around those planets gave better values for their masses, and these values combined with improved determinations of their orbits allowed more accurate calculations of the perturbations the planets produced in Mercury's orbit. The same held for Jupiter and the Earth. The use of high-speed computers to carry out these computations also played a key role. In fact, modern-day celestial mechanicians still marvel at how Le Verrier, Newcomb, and their contemporaries managed to do these calculations all by hand. This space-age celestial mechanics is carried on at such places as Lincoln Laboratory at the Massachusetts Institute of Technology (MIT), the Center for Astrophysics at Harvard University, the Jet Propulsion Laboratory in Pasadena, California, the U.S. Naval Observatory in Washington D.C., and at similar installations in Europe, Japan, and the USSR.

A compilation of a decade's worth of data (1966–76) by the MIT group gave a value for the anomalous part of Mercury's perihelion precession of 43.11 ± 0.21 arcseconds per century. Compare that with the general relativistic prediction of 42.98—complete agreement, within the experimental errors.

The Earth, for example, also experiences a perihelion precession, caused partly by planetary perturbations and partly by relativity. The relativistic effect is smaller than it is for Mercury, because the Earth is farther from the Sun, so the space curvature is smaller, the Earth's velocity is smaller, and the gravitational nonlinearity effect is smaller. On the other hand, the effect of planetary perturbations is larger, mainly because the Earth is closer to the massive Jupiter. Nevertheless, it is measurable, although with larger observational errors, and the best value is 5.0 ± 1.2 arcseconds per century, compared to the relativistic prediction of 3.8. Within the experimental errors, all would seem to be well with general relativity.

But in 1966, observations of the Sun by Robert Dicke and H. Mark Goldenberg started a vigorous debate over the validity of Einstein's perihelion prediction that continues to this day.

Is the Sun a sphere? This seemingly innocuous question has exercised relativists and solar physicists for over twenty years, ever since Dicke and Goldenberg reported that it wasn't. During the late spring and summer of 1966, Dicke and Goldenberg measured the visible shape of the Sun by making clever use of a device called an occulting disk. This is a circular disk that can be placed in front of the image of the Sun in a telescope in order to block out all or part of the Sun's image, depending on the diameter of the disk and the magnification of the telescope. In this way, one can produce an artificial eclipse of the Sun. Dicke and Goldenberg chose a disk and a magnification that left only the very edge of the Sun, called the limb, visible. If the Sun were slightly out of round, that is, fatter at its equator than at its pole, then the thin circle of light that appears around the occulting disk would appear thicker, and therefore brighter on opposite sides of a diameter drawn across the solar equator, and dimmer on opposite sides of a diameter drawn along the solar axis. This difference, if it existed, would be at most a few parts in a hundred thousand, so it could not be detected by the naked eye. Instead, it could be measured using precise light-measuring devices called photodetectors. In practice, of course, the image of the Sun in a telescope is highly distorted from a circular shape by the mirrors in the telescope, and by refraction or bending of the Sun's rays by the Earth's atmosphere, among other things. These effects had to be corrected for very carefully.

A more serious and subtle problem had to do with the Sun itself. Suppose the circle of light around the occulting disk turned out to be brighter at the solar equator than at the pole. One interpretation would be that the Sun is indeed fatter at the equator, and that we are seeing more of it there. But another interpretation might be that the Sun's image is perfectly circular, but that the Sun is intrinsically brighter at the equator than at the pole. This is actually not an unreasonable idea, because it is known that sunspots and bright regions known as faculae that are often associated with them are generally confined to the equatorial latitudes of the Sun. How can one distinguish between these two possibilities? One of the ways used by Dicke and Goldenberg was to change the effective size of the occulting disk by varying the magnification of the telescope (see figure 5.2). In this way they could measure the brightness difference with double the thickness of the visible circle, and again with triple the thickness. If the Sun were indeed circular in shape but brighter at the equator, then, as a thicker ring of the Sun is exposed, the newly exposed portion should have the same brightness difference as the first ring, so the measured brightness difference should double. At triple the thickness, the measured brightness difference should triple. On the other hand, if the Sun is uniformly bright, but fatter at the equator, then, as an additional ring is exposed, the light received from it should be uniformly bright around the circle, and the net brightness difference should be unchanged. At triple the thickness of the circle, the brightness difference should still be the same. In fact, Dicke and Goldenberg reported very little change in the brightness difference with size of the disk. This and other tests convinced them that they were indeed seeing the effects of a flattened, or oblate Sun.

Now, there is absolutely nothing wrong with such an idea. The Sun is known to rotate with a period of about twenty-seven days, and the resulting centrifugal forces will naturally cause it to bulge at the equator and contract at the poles. The same phenomenon causes the Earth to be oblate, with an equatorial diameter about 40 kilometers larger than its diameter along its rotation axis. Assuming that the Sun rotates with the same period throughout its interior as it does at its surface, theorists can estimate the expected amount of oblateness for it. The

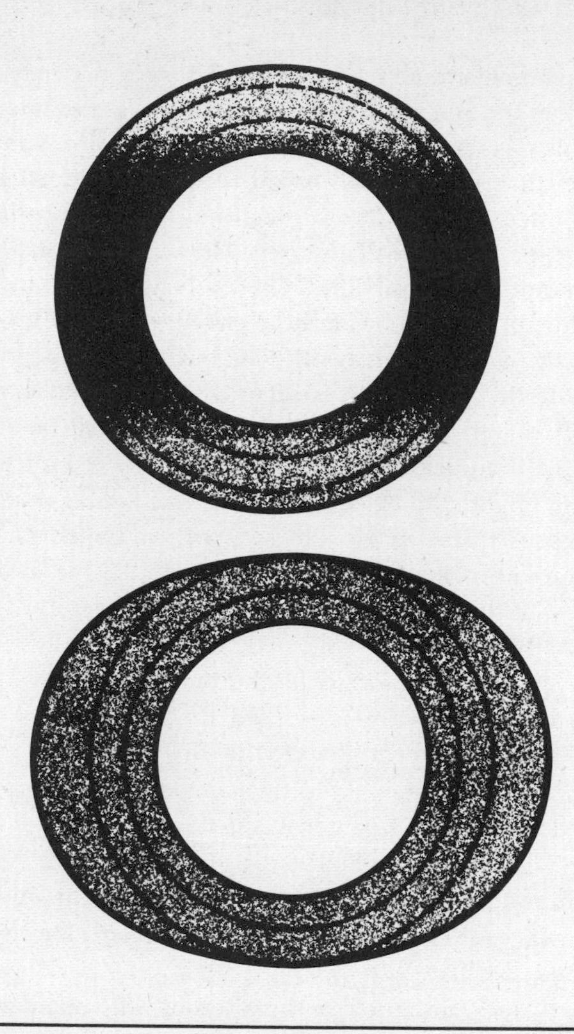

Figure 5.2 Occulting disks and the solar oblateness. Circular occulting disk is placed in front of the image of the Sun (shown in gray), and the brightness difference between pole and equator of the visible part of the Sun is measured. Apparent size of the disk relative to the Sun can be varied by altering the magnification in the telescope. In top figure, the Sun is circular, but brighter in the equatorial regions. As the disk is reduced in size, exposed regions (inside dashed circles) are also brighter in the equatorial regions, so the brightness difference increases. In bottom figure, the Sun is flattened, but of uniform brightness. As the disk is reduced, exposed regions have uniform brightness, so the brightness difference does not change.

result is a diameter difference of only 200 meters, or about 1 part in 10 million.

The trouble is that this value is nothing like what Dicke and Goldenberg found upon analyzing their brightness data. They found a diameter difference of 52 kilometers, or about 4 parts in 100,000, over 250 times larger than expected. They also quoted errors in their numbers of ±10 percent. Dicke himself proposed a possible explanation for this large value. He postulated that the interior core of the Sun, out to about one-half of its radius, is rotating 20 times more rapidly than the surface, say at a rate of once every 1.3 days. The additional centrifugal force produced by this rapid rotation would be enough to flatten the Sun in an amount consistent with their observations.

This caused a considerable stir in the solar physics community, and a mammoth number of papers was written, some in favor of the large oblateness, some in opposition. One recurring line of argument in opposition was that the Dicke-Goldenberg brightness differences could still be consistent with a nearly spherical Sun whose equator was intrinsically brighter, if their data using different-sized occulting disks were interpreted properly. Other physicists argued against the idea of the rapidly rotating core. A certain amount of excess rotation would be acceptable, they claimed, on the grounds that the outer layers of the Sun are probably slowing down because the solar wind that is constantly streaming from the solar surface carries off some angular momentum. The resulting friction between the more slowly rotating outer layers and the more rapidly rotating inner layers will gradually slow down the latter until the entire Sun rotates uniformly as a rigid body. However, this effect takes time (although exactly how long is difficult to calculate and depends on many assumptions about the solar interior), and it might not be surprising if the core had not yet slowed down to the same rate as the surface. What was difficult to accept, according to these solar physicists, was the large size of the rotation rate of the core required to account for the oblateness measurements.

Refutations of these arguments by Dicke and his supporters and counterrefutations abounded in the scientific literature. Dicke has written that responding to these critical papers took up so much time that it delayed publication of the detailed description of the experiment and of the data analysis for almost eight years, until 1974. Nevertheless, he said that he benefited greatly from being forced to consider those criticisms carefully. However, he remarked that it was amusing to notice, in contrast, that his publication of the Princeton version of the Eötvös experiment (see chapter 2), whose result was more conventional, supporting Einstein's equivalence principle, "did not cause the slightest ripple."

The Dicke-Goldenberg measurements caused an even bigger uproar in the relativity community, for if they held up, they would have disastrous consequences for Einstein's theory. The reason goes back to Newtonian gravitation. It is usually stated that the force of gravity between two bodies decreases as the inverse square of the distance between them, but strictly speaking this is only true if the bodies are spheres. If either one of the bodies is not exactly a sphere, if it is slightly oblate, for example, the force of gravity is modified in two ways. First, for an observer located at a given distance from the center of the oblate body, the force will be slightly smaller if the observer is located along the axis of the oblate body than if the observer is located in the equatorial plane; whereas, if the body is spherical, the force at a given distance will be the same no matter what the direction to the observer. Secondly, along any given direction, the force will no longer fall off exactly as the inverse square of the distance, but will have an additional small contribution that falls off as the inverse fourth power of the distance. These modifications in the force will cause perturbations in the orbits of planets such as Mercury just as do the additional forces caused by the other planets and by the modifications produced by general relativity. The larger the oblateness, the larger the additional forces, and the larger the resulting perturbations.

The most important consequence of the perturbation caused

by a solar oblateness is a further contribution to the perihelion shift of a planetary orbit. If the oblateness were as large as that suggested by the Dicke-Goldenberg measurements, the contribution to Mercury's perihelion advance would be about 3 arcseconds per century. This would be very bad for general relativity, because, starting from the observed anomalous advance of 43 arcseconds and subtracting the 3 arcseconds due to the supposed oblateness, we are left with only 40 arcseconds to attribute to relativity. But the prediction of general relativity is fixed at 43 arcseconds; it can't be fiddled with.

On the other hand, the alternative theory of gravity that Dicke and Brans had devised in 1960 predicted a perihelion shift that was automatically slightly smaller than that of general relativity. Furthermore, the prediction could be adjusted to fit the value 40 arcseconds per century. So the perihelion shift of Mercury, combined with the solar oblateness observations of Dicke and Goldenberg, appeared to cast doubt on general relativity and to favor the Brans-Dicke theory. The ferment generated by this conundrum played an important role in the rejuvenation of interest in the status and applications of general relativity that reached a peak between the years 1967 and 1975. In fact, the importance and impact of the Brans-Dicke theory on the verification of general relativity is a fascinating subject because it illustrates in a beautiful way the scientific method—the interplay between theoretical ideas and experimental tests that results in a deeper understanding of nature. (See chapter 8, which is devoted to the subject.)

As we will see, the Brans-Dicke theory is now in decline because its predictions turned out to be in disagreement with other experiments. Nevertheless, the problem of Mercury's perihelion and of the solar oblateness remains unresolved.

Dicke and Goldenberg's 1966 visual oblateness measurements have been neither retracted nor refuted. In 1974, they published a complete reanalysis of their data, and came up with the same result. Disagreement over the interpretation of the data as a true oblateness versus a difference in the intrinsic solar brightness between poles and equator continues. One obvious step toward resolution of this problem would be to repeat the observations. This was done in 1973. Interestingly, the leader of the group that made the measurements was Henry Hill, who in 1966 was an assistant professor at Princeton, associated with Dicke's group. Before leaving Princeton, he had played a major role in developing the telescope system used by Dicke and Goldenberg. For the 1973 measurements, Hill worked at an observatory specially designed for solar observations, the Santa Catalina Laboratory for Experimental Relativity by Astrometry (SCLERA), located in the Santa Catalina mountains outside Tucson, Arizona. Astrometry is a branch of astronomy involved in high-precision measurements of the positions of stars. Hill's method did not use occulting disks; rather, it used photodetectors to measure the edge of the Sun directly and thereby to determine its shape. The results implied a difference in the pole and equator diameters of only 2 kilometers, or about 1 part in a million. The experimental uncertainties were plus or minus 5 times this amount, so the results were consistent with a very small oblateness, such as that predicted by a conventional solar model. The maximum value permitted by Hill's results was also at least 5 times smaller than the Dicke-Goldenberg results. This seemed to turn the tide back toward general relativity and the conventional Sun.

However, soon the tidal waters receded from general relativity once again, the instigator being Hill himself. Working at SCLERA, Hill's group had been monitoring the vibrations of the Sun, first detected in 1976. Like the overtones of a guitar string, these fluctuating distortions of the Sun occur at well-defined frequencies, ranging from a dozen cycles per hour down to a cycle every few hours. The cause of these vibrations is still poorly understood, but by 1980 they had been well studied observationally by several groups. It turns out that many of the characteristics of these oscillations, such as the differences between the frequencies of oscillations of a similar type, are sensitive to the internal rotation state of the Sun. From detailed analyses of the data, Hill and his co-workers, as well as other researchers, claimed in 1982 that the core of the Sun is rotating more rapidly than the surface, by a factor of 6. This was not as rapid an internal rotation as Dicke had claimed, but it was enough, according to these workers, to produce a flattening of the Sun of 7 parts in a million, with experimental uncertainties of about 20 percent. True, this was 5 times smaller than the Dicke-Goldenberg value, but it was still 50 times larger than the conventional value. The contribution of such an oblateness to the perihelion shift of Mercury would be about half an arcsecond per century. Because the radar measurements of the shift agreed with general relativity to about one-fifth of an arcsecond, such a contribution was uncomfortably large, though not as disastrous as it was using the earlier oblateness values. To make matters even more complicated, later observations and analyses of solar oscillations by other workers produced values of the solar oblateness very close to the small conventional value. Dicke got back into the game in 1985, with a report that new visual oblateness measurements made during the spring and summer of 1983 yielded a value of 12 parts in a million, one-third of the 1966 value, but larger than Hill's solar oscillation value. By the middle 1980s, there was still no consensus on the size of the Sun's oblateness.

There ought to be another, independent way to measure the solar oblateness that avoids having to worry about such messy issues as the brightness of the solar surface, its internal rotation, or its vibration. Indeed, there is such a way, but it has proved to be difficult to implement. The method is to try to measure directly that additional component of the gravitational force exerted by the Sun that falls off as the inverse fourth power of the distance. This component depends directly on the oblateness of the solar matter distribution and is directly responsible for the perturbations that might produce any additional perihelion advance of Mercury. One way to do this would be to measure the perihelion advances of planets other than Mercury. The reason this would be useful is that because the contribution of the oblateness to the gravitational force falls off as the inverse fourth power of distance, it rapidly becomes negligible as one considers planets farther out from the Sun. Thus, for example, the anomalous perihelion advance of the Earth (after the perturbations of the other planets have been accounted for), would be purely due to relativity, the oblateness contribution being utterly negligible. Unfortunately, the effect of relativistic perturbations also falls off with distance, albeit not as rapidly, so that, as we have already mentioned, the predicted advance for the Earth is only 3.8 arcseconds per century. Also, the precision of the measurements of this value is only about 20 percent, not good enough to distinguish between general relativity and the Brans-Dicke theory. Venus is closer to the Sun, and therefore might have a sizable oblateness contribution as well as a larger relativistic contribution, but unfortunately, its orbit is so close to being a pure circle that it is extremely difficult to locate the perihelion point with any reliability, and as a consequence, the errors in the observed value of its anomalous precession rate are more than 50 percent.

Another way might be to search for planetary perturbations other than the perihelion shift that the oblateness term might produce. For example, the orbits of Mercury and Mars are elliptical, so that their distance from the Sun varies from a minimum at perihelion to a maximum at aphelion. Therefore, the relative contribution of the oblateness term in the gravitational force will vary from point to point in the orbit, and this will result in periodic variations in such variables as the direction and speed of the planet. These small variations can now be detected in the orbits of Mercury and Mars by radar ranging. Analysis of this radar data typically takes years and is not yet complete, but preliminary results suggest the possibility that the oblateness is closer to the conventional value than to the larger values inferred from previous observations of the Sun. No one has been able to resolve or understand the discrepancies between all these values for solar oblateness, other than to say that the observations are difficult and subject to many errors.

If it is reasonable to look for the variable perturbations of the oblateness term on an elliptical orbit, then why not go all the way? Why not send a body from the largest reasonable distances, such as the orbit of Jupiter, all the way to the smallest distance possible, the center of the Sun? The contribution of the oblateness term in the gravitational force to the motion of the body would increase by a factor of several billion, ranging from the utterly negligible far away to a measurable size close to the Sun. The relativistic effects would also vary during such a flight, increasing in size the closer the body gets to the Sun, but they have a different dependence on the distance from the Sun than do the oblateness effects, and would increase by only a factor of 100,000. Therefore, in principle, the effects of the two phenomena could be determined separately by continuous tracking of the body. As impractical as such an idea might seem, it has actually been under serious consideration by scientists for over a decade. Originally named "An Arrow to the Sun" and now referred to as "Starprobe," the mission first involves a flight to Jupiter (see figure 5.3). A swing by that large planet is necessary to kill the centrifugal forces that keep the spacecraft away from the Sun, forces that are present by virtue of its being launched from the orbiting Earth. If the craft is to hit the Sun, it cannot have much velocity around the Sun, and no rocket booster exists or is likely to exist that has enough power to cancel the Earth's orbital velocity and allow a spacecraft simply to "drop" into the Sun. Further study of the mission showed that it actually would be better to leave the craft with a small amount of angular velocity, so that instead of crashing into the Sun, it would just miss it, by about four solar radii, or one-twentieth of the radius of Mercury's orbit. Feasibility studies of such a mission suggested that the oblateness could be measured quite accurately. Even if it were as small as the conventional value of 1 part in 10 million, it could be measured with an accuracy of around 10 percent. If successful, this would settle the issue of the solar oblateness and Mercury's perihelion. A mission of this sort would also be of great benefit to solar

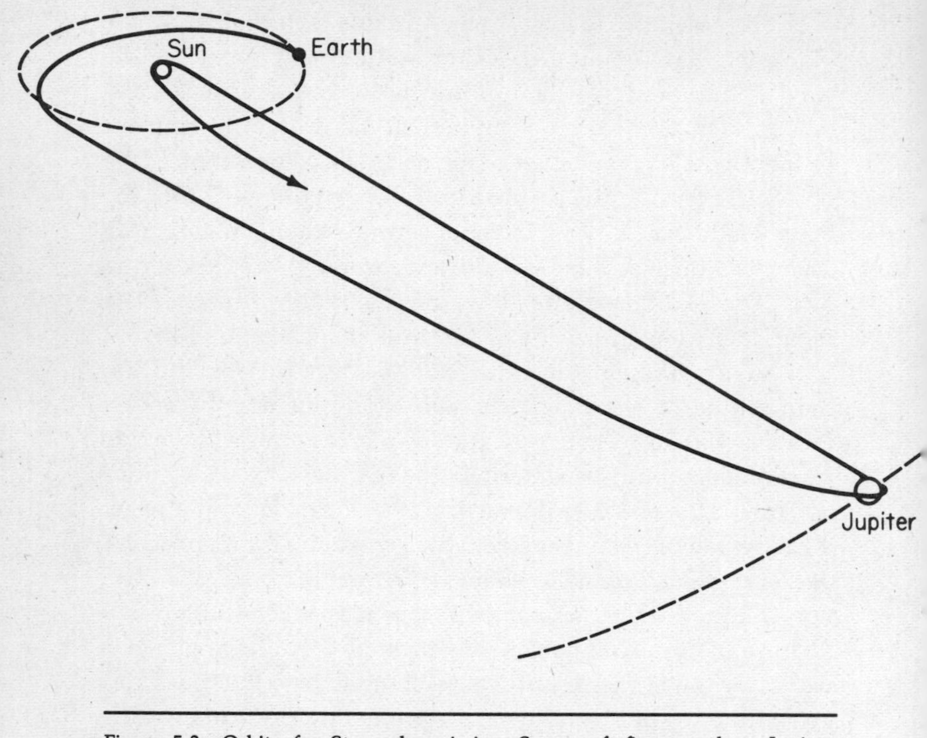

Figure 5.3 Orbit of a Starprobe mission. Spacecraft first travels to Jupiter, using that planet's gravitational field to pull it around so that it can fall almost directly toward the Sun.

physicists, because the spacecraft could make measurements of magnetic fields and the solar wind very close to the Sun, and could study the solar corona and photosphere with a level of detail impossible to achieve by any other means. Some relativists have argued that this may be one of the most important space missions of the decade.

Unfortunately, by 1978, just when relativists and solar physicists were becoming excited about the scientific potential of such a mission, NASA was entering a dry period when it proposed no new interplanetary missions, and instead appeared to focus most of its attention on the space shuttle. At present, the Starprobe mission is in limbo.

It is ironic that after seventy years, Einstein's first great success remains an open question, a source of controversy and debate. Like Einstein in 1919, many general relativists and solar physicists are confident that the issue will ultimately be decided in favor of general relativity, but confidence is not the final arbiter in science. We must await further observations, possibly from Starprobe, before we can really close this chapter.

A second irony is that much of our confidence in general relativity is now based on an experiment that Einstein never dreamed of. It is a late-comer to the pantheon of tests of general relativity, yet it has given the most precise confirmation of the theory of all. To see how this new test, called the "time-delay of light," came about, we must return to the Venus radar measurements of 1959.

6

The Time Delay of Light: Better Late Than Never

IF MERCURY was present at the birth of general relativity, then Venus and Mars participated in its coming out. Mercury's perihelion precession provided one of the earliest confirmations of the theory, yet today this test is still somewhat clouded by controversy. It was Venus that led to a new test of general relativity, not thought of by Einstein, and Mars that led to fulfillment of this test, with an accuracy unsurpassed by any other experimental test of general relativity.

The first detection of a radar echo in the September 14, 1959 radar bounce experiment with Venus opened up a new field of planetary radar astronomy. Radio contact with Mars and Mercury followed. This new program offered major improvements in our understanding of the inner planets. For example, the distortions in the radar signal caused by its reflection from the planetary surface could give information about the rotation of the planet and about its surface features, or topography. The fact that Venus rotates retrograde, that is, in the opposite sense to its orbital motion, was discovered by radar. It had previously been undetectable by optical means because of the thick cloud cover of the Venusian atmosphere.

These were truly remarkable achievements. A typical radar facility, such as Haystack, will put out a blast of radar with up to 400 kilowatts of power. However, because the radar beam spreads out in space over the hundreds of millions of kilometers that separate the planets, because the planetary surface is not a good reflector, and because its convex shape causes the reflected beam to be spread out even further, the returning echo is unbelievably weak, about one billion-trillionth of a watt. As one of the pioneers of this method wrote at the time, this is "less power than would be expended by a housefly crawling up a wall at the rate of one-millionth of a meter per year." Nevertheless, the echos can be rich in information.

In addition to shedding light on the physical properties of the planets, the echos also promised significant improvements in our knowledge of their motions. By measuring the time required for the radar signal to make its round trip, the observers could determine the distance from the Earth to the planet to about 1 kilometer, or to 1 part in 100 million. (The round-trip travel time could be converted into an effective distance by dividing by 2 and multiplying by the speed of light.) By making such distance measurements systematically during the orbits of both the Earth and the planet, the observers could obtain better determinations of the orbits of both.

As we saw in chapter 1, the early radar echo observations led to a revision in the value of the astronomical unit upwards by 93,000 kilometers. Why were such measurements important? The first reason had to do with space travel. By the early 1960s, the U.S. and Soviet space programs were in full swing, and plans were being made for possible orbiters and landers on planets such as Mars and Venus, to say nothing of the proposed manned landing on the much closer Moon. Prior to radar measurements, typical interplanetary distances could be computed to little better than 1 part in 1,000, and because the closest any planet comes to the Earth is 40 million kilometers, an error of

40,000 kilometers could mean the difference between a soft landing on a flat plane, a shattering crash on a mountain top, or an embarrassing miss of the planet altogether. Improved tracking of the planets played an integral part in interplanetary exploration. The second reason was more fundamental. Dramatically improved knowledge of planetary motions made possible improved tests of the basic laws of physics that govern the solar system.

It was this latter issue that was on the mind of Irwin Shapiro in the early 1960s. Since receiving a Ph.D. in physics from Harvard University in 1955, Shapiro had been at MIT's Lincoln Laboratory during those early radar echo experiments to Venus, working on the problem of improved determinations of the astronomical unit by radar ranging to planets. He also worked on questions of the effects of radiation from the Sun and of the Earth's magnetic field on the Earth-orbiting satellites that were being put up with increasing regularity following the Sputnik launch of 1957. He had already concluded that radar observations could be used to give improved measurements of Mercury's perihelion shift. Anything that affected the observed motion of a planet or satellite was of interest, even if it was something that influenced the radar signal used to track it.

Nevertheless, Shapiro had only a passing acquaintance with general relativity, and might not have ever considered it relevant to radar ranging had it not been for a lecture he attended in 1961 on measurements of the speed of light. Purely in passing, the speaker mentioned that according to general relativity, the speed of light is not constant. This statement puzzled Shapiro, because he had always thought that according to relativity, the speed of light should be the same in every inertial frame. He knew, of course, that general relativity predicts that light should be deflected by a gravitating body; the question was, Would its speed also be affected? It stands to reason that it might, because in the case of a prism of glass, for instance, the deflection of light as it passes through the prism is a consequence of the change in speed of the light as it passes from air to glass and from glass to air. The two phenomena, change in speed and deflection, appear to be closely related.

Einstein himself had already considered this possibility. Once he understood, from the principle of equivalence, that gravity could have an effect on light (the gravitational red shift), he attempted to construct a theory of gravity in which the speed of light would vary in the vicinity of a gravitating body. It was the equations from this specific theory that he used in 1911 to calculate the wrong (one-half) bending of light. However, Einstein did not take the next step, the one that Shapiro took. Shapiro consulted the classic general relativity textbook by Eddington and found that, according to the equations of the full general theory, the effective speed of light should indeed vary, just as it did in Einstein's earlier model. Shapiro then applied these equations to the problem of the round trip of a radar signal to a distant object. The result was remarkable: According to general relativity, the radar signal should take slightly longer to make the round trip than one would have expected on the basis of Newtonian theory and a constant speed of light. The additional delay depended on how close the signal got to the Sun. Just as in the deflection of light, the smaller the distance between the Sun and the ray at its closest approach, the larger the effect. Thus, the effect would be most noticeable when the target was on the far side of the solar system from the Earth, so that the signal would pass very near the Sun on its round trip. Such a configuration is called superior conjunction. For example, a radar signal sent from Earth to Mars at superior conjunction that just grazes the surface of the Sun suffers a delay of 250 millionths of a second (250 microseconds). Don't forget that the total round-trip travel time for such a signal is about 42 minutes! So the idea here would be to detect an additional delay of 250 microseconds on a total travel time of three-quarters of an hour. You would think this was a hopeless proposition until you realized that the distance that light travels in 250 microseconds is 75 kilometers. So the delay represents an uncertainty in the distance to the target of half of this, or 38 kilometers. If radar ranging could indeed achieve a precision in distance corresponding to a few kilometers, then perhaps this effect could be observed.

But wait a minute, this can't be right! According to the equivalence principle, the speed of light as measured in any local freely falling frame is always the same. How then can we say that the light slows down near the Sun, resulting in a delay?

A similar question might have been raised when we discussed the deflection of light. Light moves on straight lines in local freely falling frames, so how can it be deflected? In that case, the deflection came about because we had to consider more than one freely falling frame; in fact we had to consider a whole sequence of them all along the path of the light ray. No single frame could encompass the entire light path. Similarly, even though the rulers we laid out across the solar system were individually straight and parallel to their neighbors, the entire stretch of rulers was bent relative to a similar set of rulers far from the Sun.

The problem here is the distinction between local effects, effects that are observable in very small, freely falling frames, and large-scale or global effects, which cover a range of space or an interval in time large enough that the effects of curvature of space-time are important and cannot be described by a single freely falling frame. One indication of the global nature of an effect like the deflection of light was the fact that we could not detect it by looking at a single star or quasar; we always had to compare the light from one star or quasar with that from another that was farther from the Sun.

The same remarks apply to the time delay. The speed of light is indeed the same in every freely falling frame, but we are forced to consider a sequence of such frames all along the light path, and when we do so, we find that the observer at the end of the path determines that the light took longer to cover a given trajectory when it passed near the Sun than it would have had it passed farther from the Sun. Whether or not the observer uses the words "light slows down near the Sun" is purely a question of semantics. Because he never goes near the Sun to make the measurement, he can't really make such a judgment; and if he had made such a measurement in a freely falling laboratory near the Sun, he would have found the same value for the speed of light as in a freely falling laboratory far from the Sun, and might have thoroughly confused himself. All the observer can say with no fear of contradiction is that he observed a time delay that depended on how close the light ray came to the Sun. The only sense in which it can be said that the light slowed down is mathematical: in a particular mathematical representation of the equations that describe the motion of the light ray, what general relativists call a particular coordinate system, the light appears to have a variable speed. But in a different mathematical representation (a different coordinate system), this statement might be false. Nevertheless, the observable quantities, such as the net time delay, are the same no matter what representation is used. This is one of those cases in relativity where the careless use of words or phrases that are not based on observable quantities can lead to confusion or contradiction. We have already seen an example of this in our discussion in chapter 3 of which frequency, clock or signal, "really" changed in the gravitational red shift.

To avoid any such confusion, let us derive the time delay, at least qualitatively, using purely observable quantities throughout. The method will be very similar to the one we used to derive the deflection of light. As in that case, we will see that the time delay comes about partly because of the principle of equivalence, and partly because of the curvature of space.

Consider a light ray sent from one body at a particular moment of time toward a target on the far side of the solar system. Again, consider a set of freely falling laboratories, all of the same width, as measured by their occupants, all cunningly shot out from the center of the Sun in sequence, so that each one reaches the top of its trajectory just at the moment the light ray enters it. Each observer has two flash cubes and an atomic clock. One flash cube is triggered to go off at the moment the ray enters the laboratory, the other at the moment the ray leaves the laboratory, each one powerful enough that a distant observer, far from the solar system, can pick up both flashes. Each observer uses his atomic clock to determine how long it took for the light ray to pass through his laboratory, and using his measured value for the width of the laboratory, he calculates the speed of light. Every observer obtains exactly the same value for this speed. This is the principle of equivalence in action. On the other hand, the distant observer receives the pair of light flashes from each laboratory, and using an atomic clock identical to the ones used in the laboratories, he measures the time interval between each flash. For those observers in laboratories that encounter the light ray far from the Sun, the interval between flashes measured by the distant observer agrees with the interval of time measured by the laboratory observers (we will ignore the fact that the laboratories fall a short distance during the passage of the light signal through them; although that effect was important for the deflection of light, it is negligible here). However, as the distant observer receives flashes from laboratories that are closer and closer to the Sun, he notices that his measured interval between flashes is slightly longer than that recorded in each laboratory. By now we know why. It is simply the effect of the gravitational red shift; the interval between flashes in the laboratories near the Sun appears longer as received at great distances than it does locally. Being a true relativist, the distant observer avoids concluding that the light "slowed down," taking a longer time to cross the elevators. He sticks to observables: the time intervals measured by him were longer the closer the elevators were to the Sun. By simply adding up all the measured time intervals between flashes received from all the observers along the light path from the emitter to the target and back to the emitter, the observer can determine the round-trip travel time. The result is a delay of 125 microseconds for a round trip to Mars at superior conjunction. This delay is a consequence of the equivalence principle and hence of the gravitational red shift. The trouble is, that is only half the total effect! What went wrong?

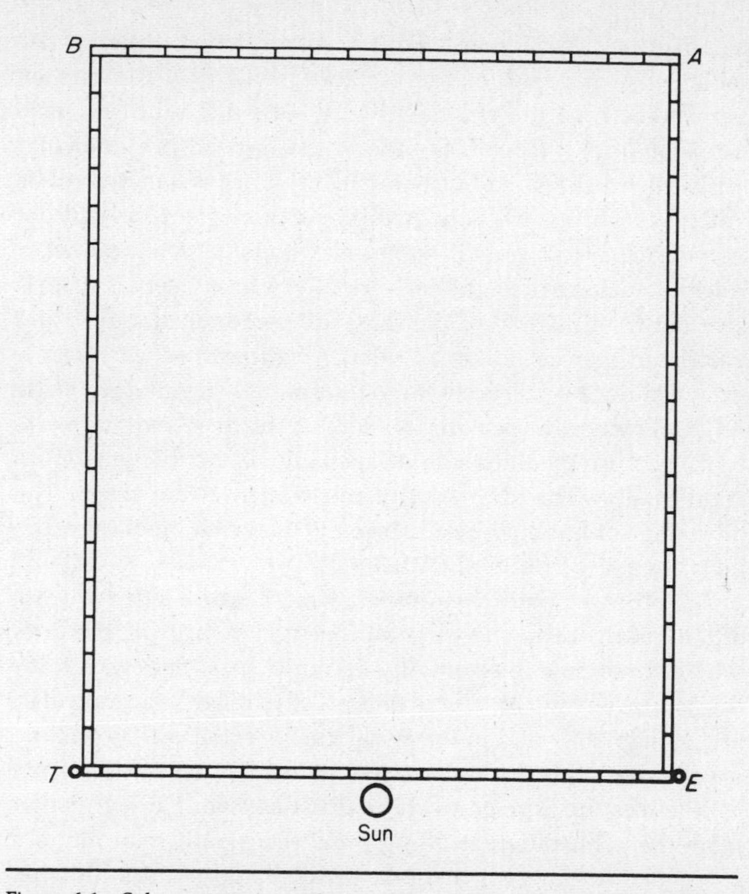

Figure 6.1 Solar-system rectangle. Line of rulers *ET* follows radar path from Earth to target, grazing the Sun. Line of rulers *EA* is perpendicular to *ET*, line *AB* is perpendicular to *EA*, and *BT* is perpendicular to *AB*. Lines *EA*, *AB*, and *BT* are all far from the Sun. Deflection of line *ET* can be ignored. The number of rulers in line *ET* is found to be more than those in *AB*.

Just as in the case of the light deflection, nothing went wrong. We're just not done. We have to take space curvature into account. To do this, we imagine the following thought experiment (see figure 6.1). An observer equipped with a large number of rulers of equal length and with perfectly square corners sets out to build a solar-system-sized rectangle, in the following manner. Starting from the emitter of the radar signal (the Earth, let's say), he sets rulers end to end, one set along the path of the radar signal heading for the target (the side *ET*), and one set perpendicular to the radar path (*EA*). The first set of rulers is extended until it reaches the target. The second perpendicular set is extended a large but basically arbitrary distance until its end *A* is far from the Sun. A third set of rulers *AB* is laid out perpendicular to this set in such a way that it parallels the set of rulers sent along the light path. Finally, a fourth set of rulers is sent out from the end of the third set, perpendicular to it, and parallel to the second set of rulers. The fourth set of rulers is extended until it meets the target, forming a gigantic rectangle *EABT*. (Practically speaking, of course, this may require a few trials, because the third side *AB* may initially be too long or too short to allow the fourth set of rulers to meet the target. The advantage of a thought experiment is that such practical problems are never insurmountable!)

The observer notes two things. First, because sides *EA*, *AB*, and *BT* of the rectangle were all constructed far from the Sun, where space-time is essentially flat, or at least flat enough for the accuracy required, all his notions of parallel, perpendicular, and straight are valid in the usual Euclidean sense. Second, he is well aware that the radar signal path *ET* is deflected slightly as it passes the Sun, so that the distance from Earth to target might be a bit different along a deflected path than along a straight path. However, a simple calculation convinces him that this effect is completely negligible. His goal now is to compare the number of rulers used in side *ET* with the number used in side *AB*, in other words, to compare the actual length of the two sides. According to Euclidean geometry, in a plane figure such as *EABT*, with right angles at each of the corners, the opposite sides should be equal in length. But to his surprise (or to his satisfaction, if he already knew about curved space), he finds that the side that passed by the Sun is slightly longer (took more rulers) than the side that stayed far from the Sun. For a rectangle whose side *ET* extends from Earth to Mars and just grazes the Sun, he finds that the extra length of the side is 19 kilometers. For a radar signal that covers this extra distance twice (once out and once on the return) the added delay in the round-trip travel time would be 125 microseconds. This is the other half of the predicted delay! It is a direct result of the curvature of space near the Sun. It is not due to the bending of the light path, which causes a negligible delay of less than one-hundredth of a microsecond.

One useful way to visualize the delay caused by space curvature is to take the two-dimensional plane formed by the light path and the Sun, and to imbed it in a fictitious three-dimensional space in such a way that the added distance measured by rulers near the Sun compared to those far from the Sun can be seen using Euclidean intuition (see figure 6.2). This is done by stretching the plane into the fictitious third dimension in the vicinity of the Sun. The picture that emerges is that of a rubber sheet with a heavy ball in the center, causing a depression. Earth and the target sit on this sheet (they too cause depressions, but they are too small to worry about). The path of the light ray is also confined to this sheet. When the Earth and the target are positioned so that the light path never passes close to the Sun, the time taken for the round trip is just what you would expect using the Euclidean distance and the speed of light. However, when the Earth and the target are at superior conjunction, and the light ray passes near the Sun, it must follow the contour of the sheet, and in moving into the depression and back out, it covers a greater distance, and is therefore delayed. Again, this picture gives only the curvature part of the time delay.

Just as in the deflection of light, this contribution of space curvature is the part that can vary from one gravitation theory to another. Any theory of gravity that is compatible with the equivalence principle predicts the first 125-microsecond part for an Earth-Mars experiment. The second part comes from space curvature, and it is purely a coincidence that general relativity predicts the same contribution from the two phenomena. As we shall see, the Brans-Dicke theory predicts slightly

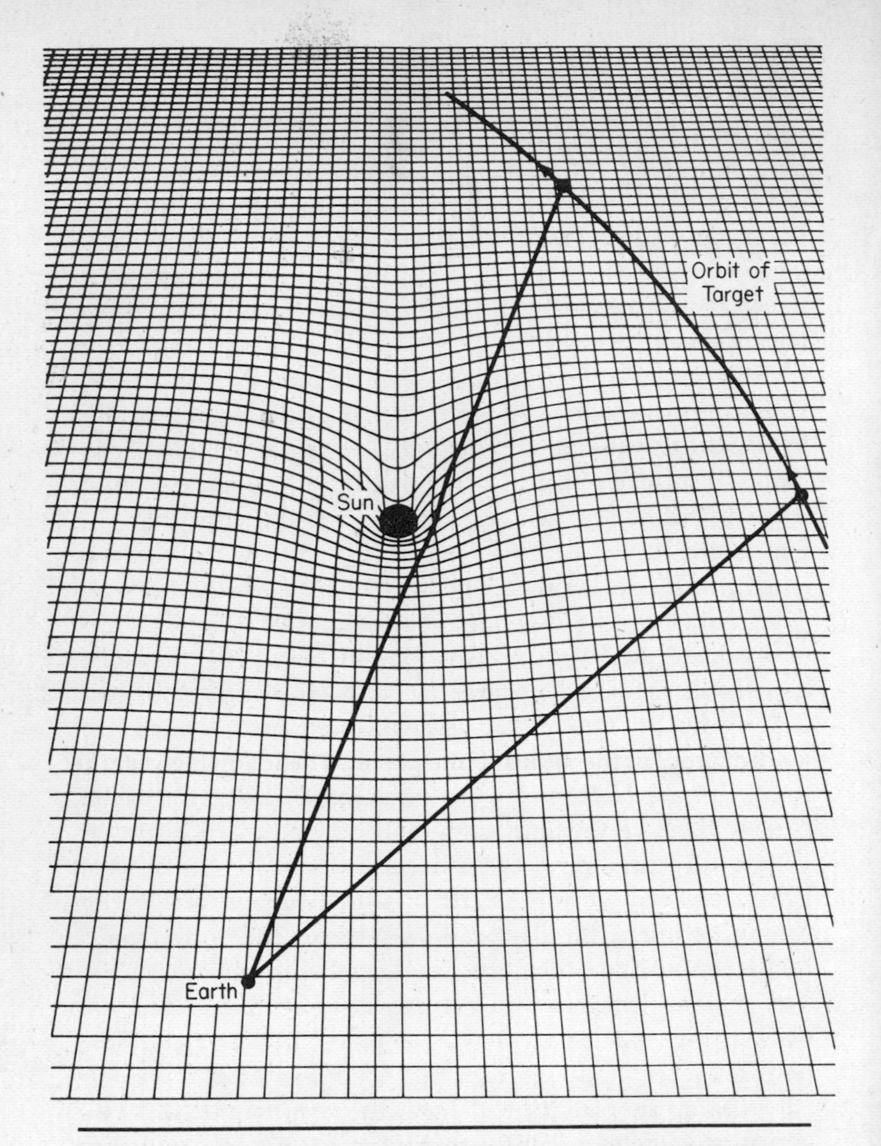

Figure 6.2 "Rubber sheet" picture of space curvature. Surface of the sheet is the two-dimensional plane formed by the light path and the Sun. Added distance near the Sun caused by curvature can be mocked up by stretching the sheet in a fictitious third dimension. Light path far from the Sun from Earth to target traverses "flat" part of sheet, so Euclidean intuition holds. When light path passes near the Sun, it must "dip into" the depression, and thus has more distance to cover.

less curvature than general relativity, and so a slightly smaller time delay.

Nowhere in the preceding discussion did we use the phrase "light slows down"; our analysis stuck to observables such as time and distance measurements. Shapiro's calculation of the time delay was based on a particular mathematical representation of the equations, so it was natural for him to speak in terms of light slowing down. As long as one doesn't ask such phrases to mean more than they are allowed to mean, no harm will be done. If one insists on asking what is really happening in the time delay, the previous discussion gives the unambiguous answer.

Shapiro did his first time-delay calculation in 1961, shortly after having his interest piqued by the lecture on the speed of light. He applied the result to the case in which the planet, such as Venus, is closest to the Earth, the configuration called inferior conjunction. The resulting delay of some 10 microseconds was interesting, but it could not be distinguished from the 1961 uncertainties in the radius of the planet. The following year, he applied the calculation to the superior conjunction case, but tossed the calculation into a drawer. It wasn't that the result, several hundred microseconds, was uninteresting; the problem was that it was impossible in 1962 to send radar to planets at superior conjunction and to detect their echos. They would be too far away for the power then available at the best radar installations. Even by the summer of 1964, the giant Arecibo radio telescope in Puerto Rico could range to Venus at inferior conjunction without difficulty, but it could just barely see Mercury at inferior conjunction. Because Mercury is closer to the Sun than Venus, it is farther from us at inferior conjunction. It would be impossible to consider ranging to any planets at superior conjunction, because they would be even farther away. So Shapiro's calculation lay in his desk for two years.

In the fall of 1964, two events caused Shapiro to retrieve his superior conjunction calculation and take it more seriously. The first was the completion of the Haystack radar antenna in

Westford, Massachusetts. The second was the birth of his son on October 30. As often happens in creative endeavors, the event in his personal life must have elevated him to a higher level of awareness or of mental activity, for soon thereafter, while describing the time-delay idea to a colleague at a party, he suddenly realized that Haystack might be able to range to Mercury at superior conjunction and provide a means to test the time-delay prediction. Shapiro decided then to write up his superior conjunction calculation for *Physical Review Letters*. The paper was submitted in the middle of November, and published under the title "Fourth Test of General Relativity" in late December, 1964. (The first three tests were the gravitational red shift, the light deflection, and the perihelion shift of Mercury, the three proposed by Einstein.)

Actually, Shapiro was not the only one in 1964 who was working on the question of the time delay. Across the continent at the Jet Propulsion Laboratory (JPL) in Pasadena, California, where the tracking of spacecraft and planets was also on everyone's mind, Duane Muhleman and Paul Reichley were also involved in a calculation of the effect of general relativity on radar propagation. Their activities raised the interesting (and sometimes controversial) question of who receives credit for major discoveries in science. Muhleman and Reichley calculated the additional delay predicted by general relativity for a radar signal from the Earth to Venus near inferior conjunction, and found delays on the order of 10 microseconds on top of round-trip times of several hundred seconds. These results were published in an internal JPL report called the *Space Programs Summary*, two weeks before Shapiro submitted his paper to *Physical Review Letters*. Shapiro was unaware of their work. Once Shapiro's paper appeared, Muhleman and Reichley subsequently confirmed his superior conjunction results in a second paper in *Space Programs Summary*, in late February 1965.

The issue is: Who should receive credit for the theoretical discovery of this important new test of general relativity? Although there is no hard and fast rule in these cases, the general consensus is that even though Muhleman and Reichley were the first into print, Shapiro should receive credit for the discovery. The reasons are twofold. First, Shapiro examined the far more important case of superior conjunction, while Muhleman and Reichley focused initially on the less interesting inferior conjunction case. Secondly, Shapiro's results were published in a widely read and easily accessible journal of physics that uses a process of refereeing, whereby a paper submitted for publication must be examined and approved by an independent anonymous physicist who is knowledgable in the subject of the paper. In questions of priority, this form of publication is normally preferred over publication in internal reports, which are usually unrefereed and less accessible to physicists at large. As a consequence, this effect is often referred to as the Shapiro time delay, while Muhleman and Reichley are credited with an independent calculation of the effect.

However, finding an effect in theory is one thing; observing it experimentally is another. In 1964, a 250-microsecond delay was in principle detectable if radar ranging to planets were indeed accurate to a few kilometers, or to tens of microseconds in travel time, but was it detectable in practice, and what accuracy could be achieved for this new test of general relativity?

The principle behind measurement of the time delay is very much the same as the principle behind measurement of the deflection of light. Just as we could not measure the deflection of a single star, we cannot detect the time delay in a single radar shot. The reason, of course, is that we cannot "turn off" the gravitational field of the Sun in order to see what the star's "true" position is or to see what the "flat space-time" round-trip travel time would have been. To get at the deflection, we had to compare the position of a star or quasar relative to other stars or quasars both when its light passed far from the Sun, and when its light passed very near the Sun. By the same token, to see the time delay, we must compare the round-trip travel time of a radar signal to the planet when the signal passes far from the Sun with that when the signal passes close to the Sun.

When the signal to the planet passes far from the Sun, the Shapiro time delay is relatively small, and the round-trip travel time is closer to being a measure of the "true" distance. This corresponds to the situation in figure 6.2 where the signal traverses a portion of space that is virtually flat. As the planet moves into superior conjunction, however, and the signal passes closer and closer to the Sun, the Shapiro time delay becomes a larger contribution to the round-trip travel time.

But there is a complication. In the case of the light deflection, we knew that the true position of the star or quasar was fixed in space to sufficient accuracy because of its large distance from us. The source itself did not move between the time we examined its position far from the Sun and when we examined it when its light grazed the Sun (it was the motion of the Earth that altered the apparent positions of the stars). Unfortunately, that is not the case here. The source, here the planet, is in orbit around the Sun, and its distance from us is continually changing. Having measured the distance to the planet by ranging when it is away from the Sun, how then can we take a radar measurement at superior conjunction and tell what part of it is Shapiro time delay and what part of it is due simply to a change in true distance between the planet and Earth?

The answer is simple in principle, though complicated in practice. Even though the radar signal may go near the Sun, the planet itself never does. Its orbit is well away from the Sun, on the order 300 million kilometers for Mars, for instance. Because of this, it always moves through a region of low space-time curvature, and maintains a relatively low velocity; therefore, the relativistic effects on its orbit are small. To the accuracy desired for a time-delay measurement, its orbit can be described quite adequately by standard Newtonian gravitational theory. Therefore, even though the planet moves during the experiment, its motion can be predicted accurately. Because of this circumstance, the time delay can be measured in four steps: (1) by ranging to the planet for a period of time when the signal stays far from the Sun, determine the parameters that describe its orbit at that time; (2) using the orbit equations of Newtonian theory, including the perturbations of all the other planets, make a prediction of its future orbit and that of the Earth, including especially the period of superior conjunction; (3) using the predicted orbit, calculate the round-trip travel times of signals to the planet assuming no Shapiro time delay; and (4) compare these predicted round-trip travel times with those actually observed during superior conjunction, attribute the difference to the Shapiro time delay, and see how well it agrees with the prediction of general relativity.

This is, of course, a gross, almost laughable oversimplification of what is done in practice. In practice, all the features that I have described above—the orbit equations of the planet, the perturbations of the other planets, the propagation of the radar signal, including the Shapiro time delay no matter where the signal travels—are built into an enormous computer program, containing more than 100,000 Fortran statements, whose output is, among other things, a statement of how well general relativity agrees with the observed travel times.

Within about a month after he submitted his paper on the time-delay effect to *Physical Review Letters,* Shapiro's colleagues at Lincoln Laboratory set out to upgrade the Laboratory's Haystack radar by increasing its power fivefold and by making other electronic improvements. This would give them the capability to get a decent echo at superior conjunction and to measure the round-trip travel times to within 10 microseconds (remember that the maximum predicted time delay was around 250 microseconds). By late 1966, the improved system was ready, just in time for the November 9 superior conjunction of Venus. Unfortunately, Venus goes through superior conjunction only about once every year and a half, so after observing Venus, they then turned the radar sights on Mercury. Because Mercury orbits the Sun almost 3 times faster than Venus, it has a superior conjunction more often, about 3 times per year, giving more opportunities to measure the time delay. Measurements were made during the January 18, May 11, and

August 24, 1967 conjunctions of Mercury. All told, over four hundred radar "observations" were used. Most of these measurements (the ones not taken near superior conjunction of either of the planets) were combined with existing optical observations of Mercury and Venus available through the U.S. Naval Observatory to accomplish the first step of our simplified description of the method, namely, to establish accurate orbits for the two planets. The remaining radar measurements centered around the superior conjunctions were then used to compare the predicted time delays with the observed time delays (because of large amounts of noise, the Venus data turned out to be not very useful). The results agreed with general relativity to within 20 percent. The first new test of Einstein's theory since 1915 was a reality.

But the story does not end there. During the summer of 1965, while Shapiro and his colleagues were busy working on the Haystack radar, a U.S. spacecraft hurtled past Mars, the first man-made object to encounter that "red planet." The spacecraft was Mariner 4, and on its way by the planet, it took twenty-one pictures and examined the Martian atmosphere using radio waves. The results both shocked and tantalized astronomers, for they indicated a rather heavily cratered, moonlike object with a cold thin atmosphere dominated by carbon dioxide, a very different picture from the Earth-like planet hoped for by some on the basis of telescopic observations. Buoyed by the success of Mariner 4, NASA in December 1965 authorized two more missions to Mars, Mariner 6 and 7 in 1969, (Mariner 5 was a Venus mission) and Mariner 8 and 9 in 1971, and planners began to think seriously about Martian landers. While these missions would bring planetary exploration to a zenith, at least temporarily, they would have crucial consequences for general relativity.

At JPL, as we have already seen, the relativistic time delay was also on people's minds, and they began to wonder if there was any way to make use of Mariner 6 and 7 to measure the time delay. There was no reason in principle why this should not be possible. Other than in size, there is no fundamental difference between a planet and a spacecraft. The orbit of the spacecraft can be determined by tracking, and its trajectory during superior conjunction can be predicted, just as for ranging to the planet, and the time delay of the radar ranging signal during superior conjunction can be measured and compared with the prediction of general relativity.

There was only one problem. Such measurements were not part of NASA's plans. The Mariner spacecraft were only supposed to photograph Mars and to otherwise study its surface and atmosphere on their way by the planet. Once this was accomplished, they were to be forgotten. Unfortunately, the Mariners were to encounter Mars when the Earth was between Mars and the Sun, exactly the opposite of superior conjunction. Because the orbit of Mars is outside that of Earth, this configuration is called opposition, and was the preferred configuration for the Martian encounter because the relative closeness of the planet at that time would ease the problems of relaying data back to Earth. There appeared to be no hope of using these spacecraft to test Einstein's theory.

The encounters with Mars were to take place in the summer of 1969, following which the two craft would leave Mars and continue to orbit the Sun in orbits similar to that of Mars, with periods of around six hundred days. Because the Earth orbits the Sun about twice as fast, then in the time taken for the spacecraft to go one-third of the way around their orbits, the Earth would go two-thirds of the way around its orbit, so that within about eight months, the Earth and the spacecraft would be at superior conjunction. What was clearly needed was an extension of the mission. An extended mission would require only that the on-board power supply of the spacecraft (provided by solar panels) be kept functional in order to operate the radar antennas and transmitters that receive and retransmit the tracking signals and to maintain attitude control so that the antennas would point toward the Earth, and that time be set aside on the NASA/JPL Deep-Space Network of radar tracking stations to make periodic tracking passes of the Mariners, perhaps through the end of 1970. In their quest for such an extension, the scientists had one foot in the door in that NASA had already approved a celestial-mechanics experiment for the original Mariner missions, whose objective was to use the accurate tracking data from the spacecraft during the Martian encounter to obtain new information on the parameters that determine the orbits of the Earth and Mars. As a result, personnel and capabilities already existed at JPL for handling and analyzing the data that would come in from an extended mission.

Nevertheless, the people at JPL decided to wait until the Mariner spacecraft proved themselves before pushing for an extended mission. Mariner 6 and 7 were launched on February 24 and March 27, 1969, and reached Mars by the end of July (see figure 6.3). Both spacecraft performed beautifully, except for one mishap. Five days before Martian encounter, a battery on board Mariner 7 exploded and left a 10-million-kilometer trail of leaked electrolyte. Fortunately, none of its other functions was affected and both craft carried out their studies of Mars without further hitch. With this success, JPL scientists and their supporters began to agitate in earnest, pointing out both the importance of an extended mission to a test of Einstein's theory, and the relatively modest cost. Still, it was by no means obvious that NASA would agree, and by November, with the original celestial-mechanics experiment scheduled to terminate, a level of panic was beginning to set in among the interested parties at JPL. Finally, on December 8, approval for the extended mission came from NASA headquarters.

The relativity experiment could begin in earnest. Between December 1969 and the end of 1970, several hundred range measurements were made to each spacecraft, with the heaviest concentration, involving almost daily measurements, around the time of each superior conjunction—on April 29, 1970 for Mariner 6 and on May 10 for Mariner 7. Neither spacecraft actually went behind the Sun. Because of the tilt of their post-Martian-encounter orbits, they both passed by the Sun slightly

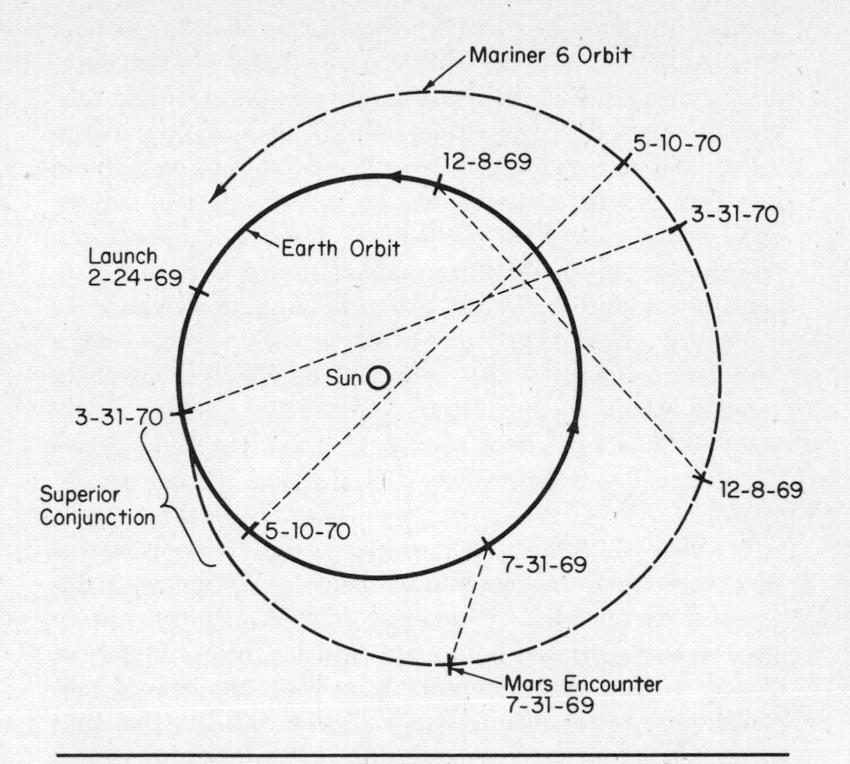

Figure 6.3 Orbit of Mariner 6. Launch Feb. 24, 1969; Mars encounter July 31, 1969, Earth and Mars close together; NASA approval of relativity experiment Dec. 8, 1969; important time-delay measurements around superior conjunction, between March 31, 1970 and May 10, 1970.

to the north, Mariner 6 about 1° away, Mariner 7 about 1.5° away. For Mariner 6, the distance of closest approach of the radar signal at superior conjunction was about 3.5 solar radii, corresponding to a Shapiro time delay of 200 microseconds out of a total round-trip travel time of 45 minutes. For Mariner 7, the radar signals came no closer than about 5.9 solar radii, giving a slightly smaller time delay of 180 microseconds. After feeding all the observations into the computer, they found that the measured delays agreed with the predictions of general relativity to within 3 percent. This was a dramatic improvement over the 20 percent figure from Venus and Mercury ranging.

Of course, the planetary radar ranging people at Lincoln Laboratory had not been idle since 1967. They had continued to make radar observations of Mercury and Venus using both the Haystack and Arecibo antennas. In fact, during late January and early February 1970, while the JPL rangers were busy getting distances to the Mariner spacecraft on their approach to superior conjunction, Venus passed through its own superior conjunction, bombarded almost twice a week by radar signals from Haystack and Arecibo. Data from that Venus conjunction, and from the numerous Mercury conjunctions between 1967 and the end of 1970, once again yielded relativistic time delays in agreement with general relativity, this time at the 5-percent level.

You might think that now the race was on—East coast versus West coast, Lincoln Laboratory versus JPL, planetary radar versus spacecraft tracking—to see who could get the most accurate test of the general relativistic time delay the quickest. As it turned out, however, the race was scratched, because both sides recognized that each approach had a crucial flaw that prevented significant improvements in accuracy. The only way to do better, it appeared, was to unite the two methods.

The flaw in planetary radar was the surface of the planet. Planetary surfaces are not perfect reflectors of radio waves, and so the returning echos are always very weak. This makes it difficult to determine their arrival times with the best accuracy. Furthermore, planets are not perfect spheres; they have mountains and valleys, and the round-trip time of a signal might depend on whether or not the center of the radar beam hit a mountain top or a valley. Suppose we are after a 1 percent measurement of a 200-microsecond time delay. This corresponds to 2 microseconds, or 300 meters in range. Mountain sizes on Mercury and Venus are on the scale of kilometers, and these topographical variations must be accounted for very accurately in order to get to 300-meter precision in the overall distance to the planet. Despite sophisticated attempts to solve the topography problem, 300 meters seemed to be the limit. Spacecraft ranging suffers neither of these difficulties. Because the spacecraft is tiny compared to a planet, there is no topography problem, and because it receives the radar signal and beams it back to Earth with a boost in power using its on-board transponder, the returning signal is strong and relatively free of noise. As a consequence, the accuracy of individual range measurements to a spacecraft at a distance of a few hundred million kilometers is better than 15 meters (or one-tenth of a microsecond in travel time)!

But spacecraft have their own critical flaw. Because they are so light, they tend to get jostled around a lot on their way through the rough neighborhood of interplanetary space. The light that the Sun emits causes a pressure, known as radiation pressure, that pushes the spacecraft significantly. (Some futuristic spacecraft have even been proposed with large reflecting solar sails to use this radiation pressure for propulsion.) The stream of ions and electrons rushing outward from the Sun, known as the solar wind, constantly buffets the craft. The steady, unchanging parts of the solar radiation pressure and the solar wind are actually well known. What is not so well known is how the spacecraft recoils from these forces. If the large solar panels that are used to derive power for the spacecraft are oriented facing the Sun, the recoil from radiation pressure and solar wind will be large, while if they are positioned sideways, the force will be less, just as for an ordinary sailboat. Because the spacecraft will alter its orientation numerous times and in complicated ways during a flight in order to communicate with the Earth, take power from the Sun, or survey stars or its destination, its response to these solar pressures is almost impossible to determine with any precision. The result is a sequence of uncertain, effectively random perturbations in the orbit of the spacecraft that in time can build up to a noticeable discrepancy between the predicted orbit and the measured orbit. Impacts by meteorites and other debris are less important because they are so rare. However, in this regard, the spacecraft itself does not help out because it is continually spitting out gas in various directions, either from its attitude control jets that allow the craft to be pointed toward the Earth or the Sun, or from the boil off of liquid nitrogen or helium used to cool various devices within the spacecraft. Newton's law of action and reaction demands that when some gas is ejected in one direction, the spacecraft must recoil in the other. A dramatic example of this effect was the battery explosion aboard Mariner 7 just before its encounter with Mars. The effect of the spewing electrolyte on the orbit was large enough to ruin any chance of using the near-Mars orbit of Mariner 7 to learn more about the mass and gravity field of the planet. Even though such effects are small in ordinary circumstances, they are essentially random and can lead to a build-up of discrepancies just like those due to radiation pressure and solar wind. Because it is necessary to predict the orbit of the spacecraft accurately near superior conjunction, these effects seriously degrade the intrinsic 15-meter accuracy claimed for spacecraft ranging. Planets, of course, are totally immune to such effects because of their enormous masses.

What we clearly need, then, is some way to combine the transponding capabilities of spacecraft with the imperturbable motions of planets. The answer is to anchor a spacecraft to a planet. One way to do this is with a planetary orbiter. A better way is with a planetary lander.

The first anchored spacecraft was Mariner 9 (Mariner 8 was lost at launch), the Martian orbiter that reached the planet in November 1971 just in time to photograph the raging dust storm that obliterated most of the planetary surface for several weeks. The next Martian superior conjunction of September 8, 1972 gave a confirmation of the Einsteinian time delay to 2 percent, only a modest improvement over the previous results, but enough to prove the power of the anchoring idea.

And then came Viking. The Viking landers on Mars were a spectacular achievement for planetary exploration, with their close-up views of the Martian surface, their analyses of the

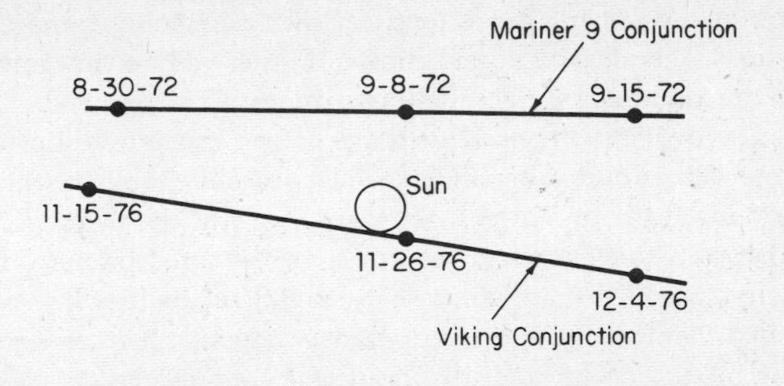

Figure 6.4 Path of Mars by the Sun during Mariner 9 and Viking experiments. Difference in paths is consequence of the tilt of Mars's orbit relative to that of Earth.

atmosphere, and their search for signs of life in the Martian soil. But to the general relativist, they were even more beautiful, for they were the perfect anchored spacecraft for the time-delay experiment. After a ten-month voyage, the first Viking spacecraft reached Mars in mid June, 1976. After several weeks of study of possible landing sites, Lander 1 was detached from the orbiter, and descended to a plain called Chryse on July 20. Eighteen days later, the second Viking reached Mars, and on September 3, Lander 2 dropped to the surface in a region called Utopia Planitia. While much of the world focused its attention on the remarkable photographs and scientific data radioed back to Earth by the landers and orbiters, the MIT group and the JPL group, now working together, were engaged in the final act of a vigorous lobbying campaign with NASA for permission to perform the relativity experiment. They had been pressing for this test almost from the moment of the Mariner 6 and 7 successes. But, as it had in the previous missions, NASA held up approval of the test until the eleventh hour. Time was running out, for the next superior conjunction of Mars was coming up soon, on November 26 (see figure 6.4). Finally, at an early September meeting of the Viking Project at JPL, Shapiro made his pitch for the time-delay test via a long-distance telephone connection from Cambridge, while someone at JPL showed his viewgraphs to the assembled panel. The relativity test was approved.

The two groups had high hopes for an accurate test of the time delay. With two landers and two orbiters all providing ranging data, they had a terrific configuration of anchored spacecraft so as to avoid the errors of random orbit perturbations. Fifteen years of progress in ranging technology had provided them with transmitters, transponders, and receivers (to say nothing of advanced computers to handle the data) that could get ranges accurate to 7.5 meters.

But they had one more thing going for them: S-band and X-band ranging. There is another source of error in time delay measurements that I have not yet mentioned, the solar corona. As we saw in chapter 4, radio waves passing through the ionized coronal gas near the Sun are deflected, and the uncertainties introduced by this additional bending ultimately limited the accuracies that could be achieved by the radio-wave light deflection experiments. Not only does the corona bend the radio-wave path, it also slows the wave down, causing an additional time delay in a signal that passes near the Sun. Out of a 250-microsecond Shapiro time delay at superior conjunction, the corona could produce an additional delay of as much as 30 microseconds. In previous time-delay experiments, this could be taken care of to some extent by modeling the corona, just as was done in many of the radio deflection measurements. However, the corona is a very turbulent phenomenon, with many large, random fluctuations in the density of the gas, and simple models are inadequate to handle these effects to the accuracy required. This is where S- and X-band come in. These terms refer to two of the standard radio frequencies used in spacecraft navigation, S-band corresponding to a frequency of about 2.3 billion cycles per second (2.3 gigahertz), and X-band corresponding to about 8.4 gigahertz. Now, the slowing down of radio waves by the solar corona depends on the frequency of the signal—the higher the frequency, the smaller the effect—in fact, the effect decreases as the inverse square of the frequency. On the other hand, the Shapiro time delay is the same no matter what the frequency of the signal. Therefore, by measuring the time delay near superior conjunction at two different frequencies, one can take the effect of the corona into account accurately. Dual-frequency ranging is so important in this regard that proponents of the time delay experiment had tried to talk NASA into adding X-band to the standard S-band system on the previous Mariner spacecraft, but without success, mainly because of the additional cost. They had tried vigorously to get X-band added to both the Viking landers and orbiters, also without success. The best they could get was an X-band downlink on the orbiters. This means that an S-band signal is sent to an orbiter, whose transponder then produces an X-band signal from it by a process of frequency multiplication, and sends both signals back to Earth. The landers would still have only S-band. Still, half a loaf was better than no loaf at all: The difference in the travel times of the S- and X-band downlinks from the orbiters could be used to estimate the effect of the corona on each round-trip radar shot, and this effect could then be subtracted from the S-band travel times to the landers, these being the more important data because the landers were more strongly anchored to the planet.

Ranging measurements were made from the moment the landers reached the Martian surface, through the November superior conjunction, and on until September 1977, when the two groups felt they had enough data to measure the Shapiro delay. A year and a half of data analysis ensued. The final result was a measured time delay in complete agreement with the prediction of general relativity, with an accuracy of 0.1 percent, or 1 part in 1,000!

The story of the Viking spacecraft that provided such a beautiful test does not end here. The landers and orbiters were designed to last for ninety days on and around Mars, but instead they lived for years. Lander 1 held out the longest, dying finally on November 20, 1982. The orbiters also operated for several years, until July 1978 and August 1980. Ranging continued to each craft right up to the end, giving a wealth of data that ultimately will be used to improve the time-delay result, because it included the next Martian superior conjunction of January 1979. But as we will see in chapter 9, this data had something important to say about whether or not gravity is getting weaker.

Although the Shapiro time delay test of Einstein's theory was a late arrival, it is still the most accurate test of the theory ever performed! But it is not the only latecomer to the pantheon of general relativity tests. Another new test came about, like the time delay, as a result of a theoretical discovery, technological advances, and the space program. To see the origins of this test, we must head west from Lincoln Laboratory, or east from JPL, into the center of the North American continent.

7

Do the Earth and the Moon Fall the Same?

BOZEMAN, Montana, 1967. From the third-floor rooftop of the physics building of Montana State University, you can see almost all there is to see. Around you is the small agricultural and ranching town settled by John Bozeman in 1864, now grown to 15,000 people. Of these, 5,000 are MSU students. Looking east, you see old U.S. Highway 10 beginning its climb up to the 6,000-foot-high Bozeman Pass. Mountains surround you on three sides: to the north, the Bridger Range; to the southeast, the Gallatin Range; and to the southwest, the Madison Range, topped by Lone Mountain and Gallatin Peak. Farther to the south, 120 highway miles, is the Old Faithful geyser of Yellowstone National Park. You turn west and look into the long Gallatin Valley, with the headwaters of the Missouri River and the Lewis and Clark caverns, named after the explorers who discovered this valley in 1806. You don't know it yet, but nestled at the foot of Lone Mountain is the future site of Big Sky, the mammoth ski, recreation, and condominium resort to be conceived and developed by TV news anchorman Chet Huntley, among others. Townspeople will decry the influx of "Easterners" (people from Minnesota and beyond), and the despoliation of the wilderness. But for now, you are in splendid isolation, a million miles from nowhere. Yet, in the world of general relativity, you are right in the middle.

Kenneth Nordtvedt should have been used to big cities: born in Chicago, an undergraduate at MIT near Boston, a Ph.D. degree at Stanford University near San Francisco, finally positions back in the Boston area, as a prestigious Junior Fellow at Harvard University and as a researcher in the Instrumentation Laboratory of MIT. But by 1965, he had developed a dislike for the life-style and politics of big cities, especially on either of the coasts, and had resolved to head for the heartland of America. When offered an assistant professorship at the small Montana State University in Bozeman, he accepted readily, and headed west to begin his academic career in earnest.

Nordtvedt's Ph.D. thesis had dealt with certain questions in the theory of solids, but solid-state physics was not his true love. His true love was gravitation theory, and upon his arrival in Bozeman, he began to think about general relativity and the motion of bodies. He considered the following question: Do massive bodies such as the Sun, planets, Moon, and so on, all fall with the same acceleration in a given gravitational field as do the balls of aluminum and platinum used in the Eötvös experiment? Obviously there was no experimental evidence one way or the other, because no one had ever tried to drop a planet or to bring the Moon into the laboratory. Nevertheless, it was an interesting question, because of a crucial difference between massive bodies and laboratory-size bodies. Laboratory-size bodies are held together by nuclear and electromagnetic forces, whereas planets are held together by their own gravity, the same force that attracts one body to another. In a fist-size ball of aluminum, for example, the amount of energy tied up in the gravitational attraction between the atoms of aluminum is less than a billion-billion-billionth of the mass-energy of the ball. On the other hand, in the Earth, because it contains much more matter, the energy of gravitational attraction or binding is a much larger fraction, just less than one-billionth of its mass-energy.

Now, laboratory Eötvös experiments, discussed in chapter 2, indicate that the nuclear and electromagnetic contributions to the energy of bodies all respond to an external gravitational field in the same way, because bodies of different structure or composition always fall with the same acceleration, regardless of whether one body might contain relatively more electromagnetic energy per unit mass than another body. Indeed, it was this universality, or independence of gravitational acceleration from the nature of the body, that allowed us to interpret gravity as a manifestation of curved space-time, rather than as a phenomenon connected with particular objects. But what about gravity itself? Could it not be that, perhaps because of some nonlinear gravity-gravity interaction between the external gravitational field and the internal gravitational field, a body that was tightly bound gravitationally might fall differently in a given external field than a body that was less tightly bound? For example, a white-dwarf star of the same mass as the Sun has a diameter of only a few thousand kilometers compared to the Sun's 1,400,000 kilometers, and because it is so compacted by gravity, its gravitational binding energy is about 1,000 times that of the Sun. If one dropped a white dwarf and the Sun side by side in some external gravitational field, would they fall with the same acceleration?

To try to answer this question, Nordtvedt devised a way of treating the motion of planetary-size bodies that would be valid in any curved space-time theory of gravity, or at least in a broad class of such theories. The equations he developed could encompass general relativity, the Brans-Dicke theory, and many others, in one fell swoop. To find the actual prediction of a chosen theory, such as general relativity, all one had to do was to specialize the equations by fixing the numerical values of certain coefficients that appeared in them. The calculations were complicated, with many, many terms in the final equation describing the acceleration of a massive body, but when all was said and done, two remarkable results emerged.

First, when the equations were specialized to general relativity, there was a tremendous cancelation of terms, and the result was that different massive bodies would have exactly the same acceleration, regardless of how tightly they were bound. Therefore, in general relativity, the acceleration of gravitationally bound bodies was predicted to be the same as that of laboratory-size bodies. In other words, if we were to drop the Earth and a ball of aluminum in the gravitational field of some distant body (keeping the ball and the Earth sufficiently far apart that we wouldn't have to worry about their mutual gravitational attraction), the two would fall at the same rate—a truly astronomical version of Galileo's Leaning Tower of Pisa demonstration. This beautiful prediction of general relativity, the equivalence of acceleration of bodies from the smallest to the largest sizes, is sometimes called the strong equivalence principle (see figure 7.1).

There was another remarkable result of Nordtvedt's calculations: In most other theories of gravity, including that of Brans and Dicke, the complete cancelation did not occur and a small difference in acceleration remained, depending on how strongly bound by internal gravity the bodies were. Therefore, even though these theories guaranteed that laboratory-size bodies fall with the same acceleration, as soon as one considered bodies with significant amounts of self-gravitational binding, the bodies would fall differently. In other words, in such theories, gravitational energy falls at a slightly different rate than other forms of energy, such as rest-mass energy, electromagnetic energy, and so on. Thus, theories such as the Brans-Dicke theory were compatible with the usual equivalence principle, but not with the strong equivalence principle.

The fact that general relativity predicted equality of acceleration of massive bodies was gratifying, but the fact that other theories did not was so unexpected that Nordtvedt worried that he had missed some crucial step. Right at this time, in the winter

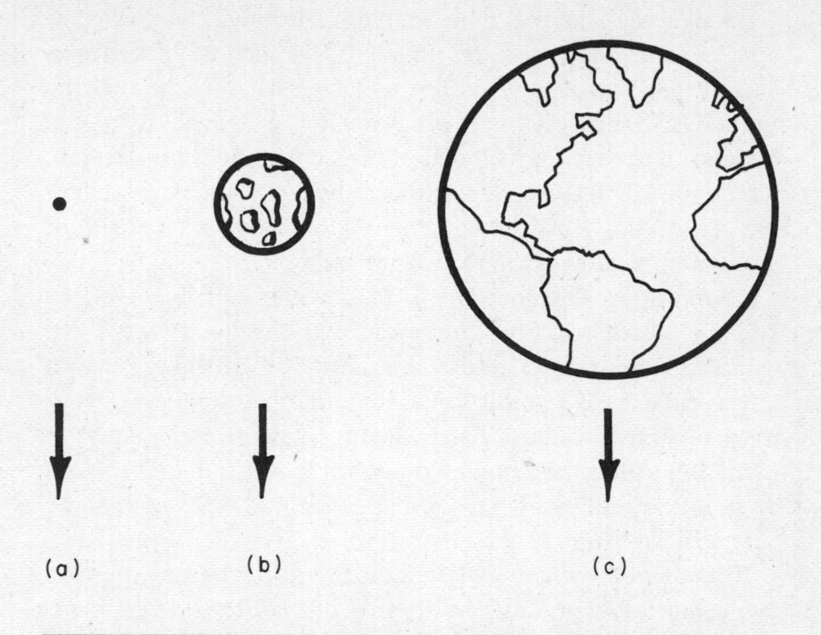

Figure 7.1 Strong equivalence principle. In general relativity, a ball of aluminum (a), the Moon (b), and the Earth (c), all fall toward a distant object with the same acceleration, even though the latter two are held together by their own internal gravity.

of 1967–68, Nordtvedt learned that Dicke himself was going to be delivering a lecture at the University of Montana at Missoula (as the only known relativist in Montana at the time, Nordtvedt had been asked to introduce the famous physicist). This was an ideal opportunity to tell Dicke about this discovery and to ask his advice. After some investigation, he learned the number of Dicke's flight to Missoula, and because the flight made an intermediate stop in Bozeman, he bought a ticket from Bozeman to Missoula on the same flight. Once in the air, Nordtvedt sought Dicke out and described his discovery. Escape for the beleaguered Dicke was unfeasible at this point. Here was a total stranger telling him that his theory violated the principle of equivalence! Dicke was noncommittal. Although he had actually vaguely considered the possibility of such a difference in acceleration some years earlier, he expressed surprise that the Brans-Dicke theory actually predicted it and said that he would have to go back to Princeton after his lecture and think about it. Undaunted by the less than enthusiastic reception his discovery had received, Nordtvedt returned to Bozeman confident enough in his own work to push ahead.

The irony in this anecdote is that, while initially skeptical, Dicke ultimately became convinced that Nordtvedt was right, that this difference in acceleration, now called the Nordtvedt effect, did exist as a prediction of his and other theories. It is the mark of a great scientist that Dicke then gave strong support to plans to put together an experiment to look for the Nordtvedt effect, an experiment whose results in the end went against the Brans-Dicke theory.

If the Nordtvedt effect existed, how could it be tested? After trying a number of ideas that turned out not to be experimentally feasible, Nordtvedt hit upon the Moon. At this time, the Surveyor program of unmanned landings on the Moon was in full swing, in preparation for the manned Apollo missions. One of the goals of the Surveyor package upon lunar landing was to deploy retroreflectors on the Moon. These retroreflectors are specially designed mirrors that can reflect a laser beam sent from Earth back in the same direction from which it came. By measuring the round-trip travel time of the laser signal, astronomers hoped to make significant improvements in their knowledge of the lunar orbit. Some relativity theorists had already begun to consider the possibility of using this improved knowledge to look for relativistic effects in the Moon's orbit, but the preliminary calculations were not very encouraging.

Nordtvedt asked, If this difference in acceleration for massive bodies exists, as it does in the Brans-Dicke theory, how would it affect the motion of the Moon? Consider the acceleration of the Earth and the Moon in the field of the Sun (see figure 7.2). The gravitational binding energy per unit mass of the Moon is about

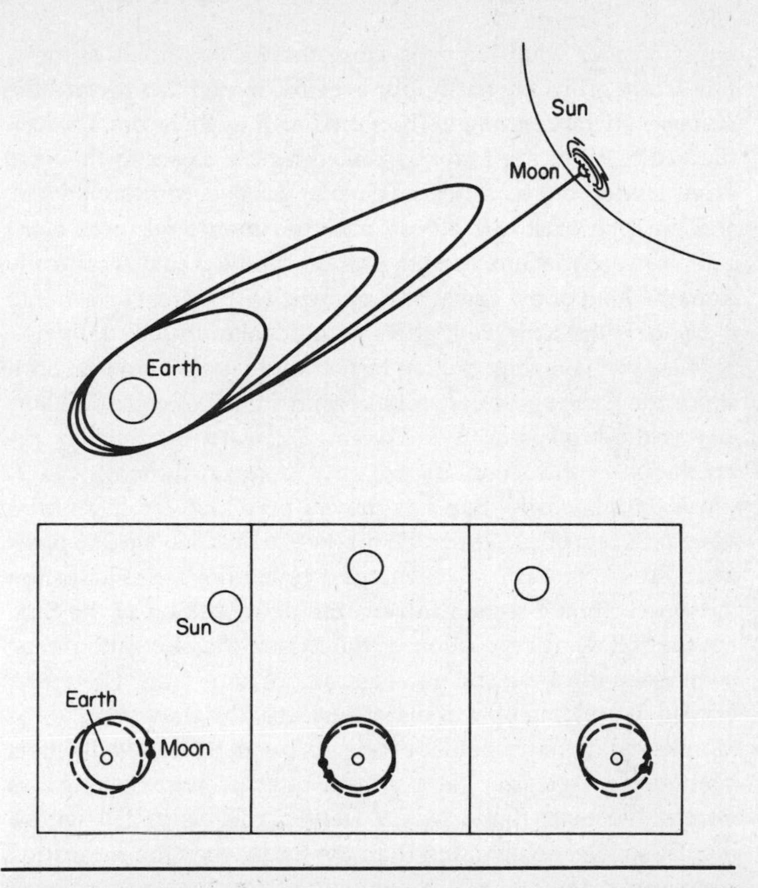

Figure 7.2 The Nordtvedt effect: Lunar catastrophe or orbit perturbation? If the Moon fell with larger acceleration than the Earth toward the Sun, its orbit would become progressively more elongated until it was pulled into the Sun. But because the Earth-Sun orientation is changing because of the Earth's orbital motion, the elongation never builds up and instead merely produces a shifted orbit that points always toward the Sun (solid curves). In general relativity, the elongation does not occur at all (dashed curves).

one twenty-fifth that of the Earth, so they could in principle fall with different accelerations because the Earth is more tightly bound than the Moon. Suppose, for the sake of argument, that the Moon falls with a slightly larger acceleration than the Earth (whether it is larger or smaller depends on the theory of gravity). Consider what happens from the Earth's point of view. The Moon orbits the Earth, but is being accelerated toward the Sun slightly more strongly than the Earth is; therefore, on each succeeding orbit the Moon is pulled a little closer to the Sun. What started out as a nearly circular orbit becomes elliptical, and on each orbit, the ellipse becomes more and more elongated toward the Sun, until the Moon is pulled catastrophically from the hold of the Earth, and plunges with a great splash into the Sun. Is the Nordtvedt effect a lunar calamity? Actually not, because we have forgotten an important fact: the Sun is in orbit about the Earth (as seen from a certain frame, of course). Thus, just as the Moon's orbit is elongated toward the Sun on one revolution of the Moon, by the next lunar revolution, twenty-seven days later, the Sun has moved by about 27° in its orbit (the Sun's rate of revolution about the Earth is 360° in 365 days, or about 1° per day), so on the next revolution, the elongation must occur in a direction toward the new position of the Sun. On the following revolution of the Moon, the elongation must be directed toward a still newer position, and so on. Therefore, instead of building up to a disastrous size, the elongation of the Moon's orbit that would be caused by the Nordtvedt effect maintains a fixed size, but is always oriented with its long axis toward the Sun. If the Moon were predicted to fall with a slightly smaller acceleration than the Earth, then the elongation would be in the opposite direction, with its long axis directed away from the Sun. If, as in general relativity, the two fell with the same acceleration, there would be no predicted elongation of this sort.

The crucial question is how large this effect might be. When Nordtvedt put in all the numbers, he found that the size of the elongation in the Brans-Dicke theory, for instance, could be as large as 1.3 meters, or about 4 feet. In general relativity, of course, the effect was zero.

As strange as it may seem at first glance, this was very exciting. Even though a 4-foot effect in the Earth-Moon distance sounds ridiculously small, in fact it was easily accessible using the proposed laser ranging to the lunar retroflectors. It was expected that the method would ultimately be able to determine the distance to the Moon to precisions on the order of 30 centimeters (about 1 foot), and so a test of the existence of the Nordtvedt effect might well be feasible. Soon after the publication of Nordtvedt's calculations, the problem of testing general relativity by looking for the Nordtvedt effect was adopted as one of the major goals of the lunar laser-ranging program.

The lunar laser-ranging program actually has its roots in the late 1950s, and like so many pioneering ideas in this subject, originated with Robert Dicke. At that time, Dicke was interested in the question of whether the gravitational constant was really a constant of nature, or whether it could change its value with time, for example, as the universe evolves (see chapters 8 and 9). One of the consequences of a gradually changing gravitational constant is that the orbit of a satellite or planet would systematically increase or decrease in diameter, depending on whether the constant of gravity was decreasing or increasing, respectively. One of Dicke's ideas was to send optical searchlight pulses to a retroreflector on an artificial earth satellite, in order to search for the effect of such changes on the position of the satellite on the sky. The development of pulsed ruby lasers and the rapid advance of the space program in the early 1960s made it possible to consider seriously making precise distance measurements directly, and using the Moon rather than a satellite. The use of lasers in the visible part of the spectrum is important because the shorter wavelength of the light allows a shorter, more well-defined pulse of light to be emitted than would be the case for the longer wavelength radar waves. The length of the pulse gives a rough measure of the maximum error that could be made in determining the round-trip travel time of the pulse. Ultimately, the ruby lasers could emit pulses with durations of tens of billionths of a second. This corresponds to a maximum distance error of about 150 centimeters, but in practice, the error could be made 10 times smaller.

Besides making it possible to look for the effect of a variable gravitational constant on the lunar orbit, the highly accurate range measurements from such a program would have many other beneficial results. They would yield a greatly improved lunar orbit and allow detailed study of the lunar librations, the angular oscillations about the Moon's center of mass that cause the face it presents to us to be not always exactly the same. They would give precise locations for the stations on Earth from which range measurements were made, and could thereby shed light on such questions as the wobble of the Earth's rotation axis and continental drift through their effects on the station locations.

But the whole idea nearly died in the fall of 1968, when the Surveyor program of unmanned lunar landers was suddenly shut down in preparation for the Apollo manned landings on the Moon, before any retroreflectors could be placed on the lunar surface. Furthermore, there appeared little hope of having a retroreflector deployed on one of the early Apollo missions, because there was no more room on the Apollo Lunar Surface Experiments Package (ALSEP) that had been designed. But then a lucky thing happened. The astronauts got too busy! It was decided by NASA that the workload for the astronauts on the first manned landing by Apollo 11 would be too heavy to deploy the complex ALSEP. By comparison, deploying the retroreflector was a breeze. The astronaut had merely to walk a distance from the lander, far enough to insure that takeoff back to the orbiter would not cover it with dust, plunk the thing down on the lunar surface, and point it facing the Earth. The device was completely passive, with no batteries or moving parts, so it was highly reliable, and furthermore its scientific objectives were important. NASA replaced ALSEP with a retroreflector for Apollo 11, to the delight of the lunar laser-ranging team, and to the presumed outrage of the ALSEP team. On July 21, 1969, the "small step" taken by Neil Armstrong onto the lunar surface also represented, with his deployment of the retroreflector, a giant step for experimental gravity. Subsequently, four other retroreflectors have been placed on the

Moon, two U.S. devices, deployed during Apollo 14 and 15, and two French-built reflectors, deposited during the Soviet unmanned missions Luna 17 and 21.

Within a week and a half of deployment of the Apollo 11 retroreflector, astronomers at Lick Observatory in California had succeeded in bouncing laser pulses off it, and measuring the round-trip travel time to a precision corresponding to several meters. By October 1969, the main laser ranging activity was centered at McDonald Observatory in Texas, and has continued there to this day. Some laser ranging is also done at observatories in France, the USSR, Japan, Australia, and recently at the U.S. observatory at Haleakala, Hawaii.

A typical laser ranging run consists of between 50 and 300 laser pulses, fired over a period of from 5 to 20 minutes. The approximate round-trip travel time for a pulse is 2.5 seconds, and by December 1971, it was possible to measure these times reliably with a precision of one-billionth of a second. This corresponds to a precision in the distance between the observatory and the retroreflector of around 15 centimeters. By May 1975, more than 1,500 successful runs had been logged, most of them with precisions on the order of 15 centimeters, and it was time to see if the Nordtvedt effect really existed. The orbit of the Moon is actually very complicated, as I have mentioned before, perturbed strongly as it is by the Sun and by the other planets, but the tremendous improvement in the knowledge of the lunar orbit provided by lunar laser ranging, combined with improved determinations of planetary orbits by radar ranging and spacecraft data, make it possible to take into account any deviations in the lunar orbit that might look like the Nordtvedt effect, but that are caused by ordinary, well-understood perturbations. After these effects are accounted for, any residual deviation could be attributed to the Nordtvedt effect. Two independent analyses of the range data were carried out, one by a collaboration of seventeen scientists from nine institutions that included Dicke and three of the pioneers of the laser ranging program, J. Derral Mulholland, Carroll O. Alley, and Peter L. Bender, the other by a group headed by Irwin Shapiro at MIT. Their results showed absolutely no evidence for the Nordtvedt effect, to a precision of 30 centimeters. When we recall that the prediction of the Brans-Dicke theory was as large as 130 centimeters, we see that the evidence goes rather strongly against that theory. On the other hand, general relativity predicts no effect, in agreement with the experimental result.

Another interesting way of expressing the results of the lunar laser-ranging experiment is to say that they verify that the Moon and the Earth fall toward the Sun with the same acceleration to a precision of about 1 part in a hundred billion. This is comparable to the precision obtained in the Princeton and Moscow Eötvös experiments on laboratory-size bodies.

In a way, the lunar laser-ranging experiment brings us full circle. The equality of acceleration of laboratory-size bodies was a foundation of the concept of curved space-time, and general relativity was built upon that foundation. However, it was by no means obvious that general relativity should predict the same equivalence for massive, gravitationally bound bodies. In the years following Nordtvedt's discovery that this was true for planetary-size bodies in which the internal gravitational binding energy was significant, but still small compared to the total mass-energy, several theoretical calculations using general relativity showed that it was also true for strongly bound bodies such as neutron stars, and even for the most tightly bound of bodies, the black hole. In other words, according to general relativity, a black hole and a ball of aluminum fall with the same acceleration. The strong equivalence principle appears to be a property of general relativity, but not of other theories such as the Brans-Dicke theory, and is another of the products of the elegance and simplicity of Einstein's theory. Einstein probably would say "of course!," and according to this lunar version of the Eötvös experiment, it seems that nature agrees.

1. Millstone Hill radar antenna, located in northern Massachusetts. Operated by the Massachusetts Institute of Technology, it was used in the early radar echo observations of Venus. (Reprinted by permission of Lincoln Laboratory, Massachusetts Institute of Technology, Lexington, Mass.)

2. Astronaut Sally Ride experiences the Equivalence Principle in action by being weightless in the Space Shuttle. (National Aeronautics and Space Administration)

3. Launch of Scout D rocket carrying a hydrogen maser clock, June 18, 1976. The clock is protected during launch by the cap covering the tip of the rocket. Comparing the rate of the rocket clock with that of a ground clock tested the gravitational redshift and time dilation. (National Aeronautics and Space Administration)

4. Carrying a portable atomic clock onto an airplane. This clock is similar to the one used by Hafele and Keating in the "Jet-Lagged Clocks" experiment. (United States Naval Observatory official photograph)

5. Radio interferometer at Green Bank, West Virginia, operated by the National Radio Astronomy Observatory. These three radio telescopes, together with a fourth one 35 km away, were used in the 1974 and 1975 light deflection measurements using the quasars 0111 + 02, 0116 + 08, and 0119 + 11. (Courtesy of National Radio Astronomy Observatory, operated by Associated Universities, Inc. under contract with the National Science Foundation)

6. Cutaway view of the Haystack radar antenna in Westford, Massachusetts, used by Irwin Shapiro and colleagues in 1967 to measure the radar time delay in signals sent to Mercury. (Reprinted by permission of Lincoln Laboratory, Massachusetts Institute of Technology, Lexington, Mass.)

7. Mock-up of a Viking lander on Mars. The antenna on the top of the lander is the S-band antenna used for ranging to the Earth. (National Aeronautics and Space Administration)

8. Apollo 14 laser retroreflector at the Fra Mauro landing site. Each circular hole in the face of the device contains a recessed reflector made of fused silica that reflects a light beam in the same direction from which it came. (National Aeronautics and Space Administration)

9. Laser shot to the Moon from the University of Texas's McDonald Observatory on Mount Locke. (Copyright the University of Texas McDonald Observatory)

10. Arecibo radio telescope in Puerto Rico. Hulse and Taylor used this telescope to discover and analyse the binary pulsar. The "dish" of the antenna is formed by the natural bowl, while the suspended superstructure can be moved to allow the telescope to be pointed. (Courtesy Arecibo Observatory, part of the National Astronomy and Ionosphere Center, which is operated by Cornell University under contract with the National Science Foundation)

11. Quartz sphere to be used in the Stanford gyroscope experiment. About 4 cm in diameter, the ball is perfectly spherical and uniform in density to about one part in 10 million. (Photo by Michael Freeman)

8

The Rise and Fall of the Brans-Dicke Theory

IF ANY one person could be singled out as being a force behind the resurgence of experimental relativity in the past twenty-five years, it must be Robert H. Dicke. Not only has he played a key role in devising and performing some of the most important experimental tests of the nature of the gravitational interaction, but he has also been responsible for many of the important theoretical insights that have profoundly influenced the way both theorists and experimentalists have approached this subject.

Yet, ironically, he has often been on the other side of the fence from the establishment, with theoretical ideas or experimental results that contradicted the conventional wisdom. As we saw in chapter 5, his measurements of the visual oblateness of the Sun in an amount over 100 times larger than standard solar theory demanded pitted him against a majority of solar physicists and general relativists, and caused a controversy that remains largely unresolved. And his development of the scalar-tensor theory of gravity with Carl Brans produced the first serious challenge to the supremacy of general relativity, generating along the way a sizable number of skeptics, and a not insignificant number of believers. But it is characteristic of Dicke that, at the same time that he was promoting the Brans-Dicke theory as a viable alternative to general relativity, he was helping to promote and bring to fruition some of the experiments that would cause its downfall. For no matter how elegant or intellectually satisfying a theory might be, it ultimately must face the test of experimentation. Dicke never shied away from this confrontation between theory and observation; in fact, he relished it, being at heart one of the great experimental physicists of our time. There are many illustrations in the history of physics of the fundamental interplay between theory and observation, but one of the most interesting is the rise and fall of the Brans-Dicke theory of gravitation.

Born in St. Louis in 1916, Dicke attended Princeton University and the University of Rochester, receiving a Ph.D. in 1941. After service at the MIT Radiation Laboratory during World War II, he returned to Princeton permanently as a faculty member in 1946. His approach to physics, whether experimental or theoretical, was to make use of general principles, such as symmetry considerations or conservation laws, to get to the heart of the problem under study, or to find a better or more elegant experimental technique. Dicke made experimental discoveries or elucidated theoretical principles that led to, among other things, the lock-in amplifier, the gas-cell atomic clock, the microwave radiometer, the laser, and the maser. He was involved, directly or indirectly, in so many discoveries for which Nobel Prizes were awarded, that many physicists regard it as a mystery (and some as a scandal) that he has yet to receive that honor. In 1964, Dicke wondered whether there might be a background of cosmic radiation left over from the big bang, and he set his Princeton research group to work on theoretical calculations to understand it and on an experiment to detect it using a microwave antenna. But in the spring of 1965, before they could complete the measurements, they learned that Arno Penzias and Robert Wilson of Bell Telephone Laboratories had accidentally detected the background radiation during the course of improving their own microwave antenna. Penzias and Wilson shared a Nobel Prize for the discovery.

Nowhere was Dicke's "elegant" attitude toward physics more useful than in his work in gravitational experiment, to which he turned around 1960. As we have seen in several places in this book, his elegant solutions to difficult experimental problems led to important new results, such as in the Eötvös experiment and the solar oblateness measurements.

But around the same time, Dicke's thoughts also began to turn toward the possibility of a new approach to gravitational theory. Over the years prior to and following the publication of the general theory of relativity, numerous alternative theories of gravity had been put forward. For the most part these theories were motivated by a desire to avoid the curved space-time aspects of general relativity. What they did instead was to retain the flat space-time of special relativity, whose experimental foundation was much more solid, and to treat gravity as a field in flat space-time, just as one treated electromagnetism using fields. But either these theories ran afoul of those experimental results that existed, such as the perihelion shift of Mercury (before solar oblateness) or the deflection of light, or they were so ugly, contrived, or complicated that no one could take them seriously. By the late 1950s, no single theory stood out as a vigorous competitor to general relativity.

Dicke, on the other hand, was perfectly comfortable with the notion of curved space-time. He agreed that it was a natural consequence of the principle of equivalence and felt that it should be retained as the basis for gravitation theory. What he had in mind was something else.

The something else was Mach's principle. This principle is actually a loose collection of thoughts about the nature of inertia and gravity that have been attributed to the nineteenth-century philosopher Ernst Mach (1838–1916). In reality, Mach never formulated anything resembling a lofty principle, and would probably not recognize some of the formulations of "Mach's principle" that have been used by physicists and philosophers during the twentieth century. Einstein himself was a devotee and promoter of what he called Mach's principle, yet in later years his own enthusiasm for it declined. Nevertheless, Mach did expound upon a basic notion or point of view that does run as a more or less common thread through most versions of the principle. The notion is this: The inertial and gravitational properties of matter are in some sense linked to the existence of the rest of the matter in the universe. A simple example, known as Newton's bucket, will illustrate this idea (see figure 8.1).

Fill a bucket with water, place it on a turntable, and start the turntable spinning. As the bucket starts to spin, the water doesn't do much of anything at first, but eventually, the friction between the water and the walls of the bucket causes the water to spin along with the bucket. As a consequence of this, the surface of the water becomes concave and the water begins to climb up the sides of the bucket, and a depression forms in the center. Quite naturally, we attribute this behavior to centrifugal forces pushing the water away from the rotation axis. When the turntable is stopped, eventually the water too slows down and returns to its initial state with a flat surface. As mundane and commonplace as this simple observation is, it has led to some of the most intriguing and vexing philosophical questions. One question is, How does the water know that it is rotating and should have a concave surface instead of a flat one? If we truly abhor the concept of absolute space, as relativity in either its Newtonian or Einsteinian forms teaches us, we cannot answer that the water knows that it is rotating relative to absolute, nonrotating space. With respect to what then? The best we can do is to answer that somehow the water knows that it is rotating relative to the distant stars and galaxies. As reasonable as this sounds, it does beg two questions. Suppose we performed this bucket experiment in an otherwise completely empty universe. With nothing to which to refer its state of motion, would the water know what to do as the turntable spun? Would its surface become concave or stay flat? That is the first question, to which

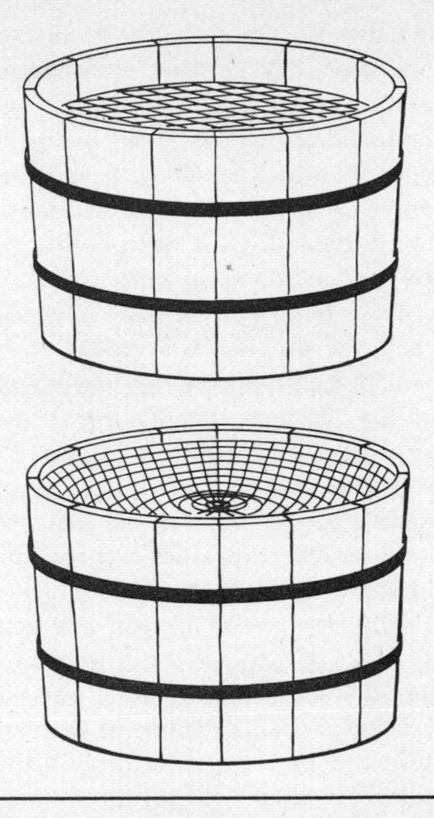

Figure 8.1 Newton's bucket. In top figure, the bucket is not rotating and the surface of the water is flat. In bottom figure, the bucket is rotating and the surface of the water is concave. Would the surface be concave if the bucket were "stationary" and the universe rotated around it?

there is no really satisfactory answer based on physics. Up to a point, of course, this question is irrelevant, because we don't live in an empty universe anyway. The second question is somewhat more meaningful: Suppose we leave the bucket at rest, and let the entire universe rotate around it with the same rotation rate as the bucket had in the previous experiment, but in the opposite sense. Would the water become concave as before? If only the rotation of the bucket relative to the distant matter in the universe is important, then the two experiments should give the same concave shape for the water's surface. In other words, it should not matter whether we say that the universe is nonrotating and the bucket is rotating, or that the bucket is nonrotating and the universe is rotating. Only the rotation of one relative to the other is relevant. Such an outcome would be compatible with Mach's principle. In principle, general relativity was capable of answering this second question, but by 1960, there had been no progress in doing so (mathematically, the problem is formidable). It wasn't until 1966 that relativity theorists were able to give a partial theoretical answer to the question, suggesting that, indeed, the two formulations of Newton's bucket experiment would agree. Ambitious experimentalists are even planning an experiment, using gyroscopes in space, to test this idea that rotating distant matter can affect the local properties of space-time. This will be one of the subjects of chapter 11.

Thus, in 1960, it was not at all clear that general relativity incorporated Mach's principle in the sense we have described, and this troubled Dicke. Another issue also concerned him. Whereas the rotating bucket experiments we have talked about dealt with the inertial properties of matter, Dicke was also concerned about the gravitational properties of matter. If you want to take the Machian viewpoint seriously, then you might argue that such a gravitational property as the size of the gravitational force between two bodies of given masses at a given separation should be somehow determined by or related to the distribution of distant matter in the universe. Now, the size of the force is set by a constant of nature known as the Newtonian gravitational constant, G, whose value is 0.0000000667 (centimeters)^3/grams (seconds)^2. Can this value be related to distant matter? Dicke noted that if you take the radius of the visible universe, given roughly by the distance that light could have traveled to reach us since the big bang, roughly 10 billion light years, multiply it by the square of the speed of light (300,000 kilometers per second), and divide by the mass of the visible universe, given by the average observed matter density (about 200 grams in every cube a million kilometers on a side) times the volume of a sphere of radius 10 billion light years, you come up with a number with exactly the same units as G and within a factor of 100 of the actual numerical value. Being only within a factor of 100 may not seem that close, but it really is when you consider that the individual numbers that went into the calculation were extremely large, so that there was no obvious reason why all these large numbers should cancel one another to give an answer so close to the true value of G. Was this another manifestation of Mach's principle? Could it be that this apparent numerical relationship between G and the mass and radius of the universe was more than just a coincidence, but instead was a law of nature? Could the value of G itself be determined by the distribution of matter in the universe? If this were true, then an immediate consequence would be that as the universe evolves and its density and radius change, the value of G could change with time. This idea is what first motivated Dicke to consider experiments that could detect such a variation in G and led him to work toward what ultimately became the lunar laser-ranging program. I will have more to say about numerical coincidences and a varying G in chapter 9.

However, general relativity was unambiguous on this point: G had to be a fixed unalterable constant of nature. There was no mechanism within general relativity for G to depend on the distribution of matter in the universe. Therefore, if Dicke wished to incorporate such an idea into gravitation, an alternative theory would be needed.

If the gravitational constant is to vary, it must be a function of space and time, in other words, at each point in space-time, a single number is to be assigned, and that number goes into the value of G. A function that does this is called a scalar field. Beginning then with the standard mathematics of curved space-time of general relativity, Dicke and his Princeton graduate student Carl H. Brans added to the mathematics a scalar field. The result was what they termed a scalar-tensor theory of gravity, tensor referring to the mathematical variable associated with curved space-time called the metric tensor. Mathematically, the theory was very much like general relativity, with some of the equations modified by the presence of the scalar field. The value of the scalar field at any given point in space and time was determined by the distribution of matter everywhere, both in the neighborhood of the point in question and in the distant universe. This value of the scalar then determined the effective, measurable value of G, thereby fulfilling Dicke's desire to incorporate a form of Mach's principle into gravity. As the universe evolves in time, the scalar field at any fixed point in space should change its value with time, and consequently, the value of G should change with time. This aspect of the theory was very satisfying.

Another aspect of the theory was less satisfying. In order to introduce the scalar field into the mathematics of curved space-time, Brans and Dicke had to introduce some uncertainty. That uncertainty took the form of an adjustable numerical constant they denoted by the Greek letter omega (ω) whose value determined how dominant a role the curvature of space-time was to play compared to that of the scalar field. The larger the value of ω, the more dominant the curvature, and the smaller the effect of the scalar. The smaller the value of ω, the larger the effect of the scalar. In fact, if one selected progressively larger and larger values for ω, the theory became indistinguishable from general relativity. Unfortunately, there was no way to come up with a specific value for ω, although in a moment we will see that Brans and Dicke were particularly attracted to a value around 7.

Except for this arbitrariness in ω, the scalar-tensor theory was every bit as valid mathematically as general relativity, and was capable of making detailed predictions for the outcomes of experiments. The predictions would depend on the chosen value of ω, of course, but that aside, the theory could do anything general relativity could do. But now the theory had to pay a price for its ability to incorporate Mach's principle into the value of G. Because its equations were slightly different than those of general relativity, its predictions would be slightly different. For example, it predicted a deflection of light by the Sun that was slightly smaller than that predicted by general relativity, 7 percent smaller for a value of ω of 5, 4 percent smaller for ω of 10, 0.5 percent smaller for ω of 100, and so on. As ω was chosen to be larger and larger, the predictions of the two theories coincided. The same percentage differences were predicted for the Shapiro time delay. The predicted perihelion shift was also smaller than that of general relativity, 10 percent smaller for ω of 5, 6 percent smaller for ω of 10, 0.7 percent smaller for ω of 100, and so on. On the other hand, because the theory incorporated curved space-time in the same manner as general relativity, it automatically agreed with the equivalence principle for laboratory bodies (Eötvös experiment) and with the gravitational red shift.

In 1960, the new theory did not contradict any known experiments. As we have seen, optical measurements of the deflection of light had sufficiently large experimental errors, at the 10 percent level, that they were compatible with the theory. The time delay did not yet exist as a possible test, let alone as a test with actual results. And although the theory predicted a perihelion shift for Mercury that was rather low compared to the observed shift, the disagreement was not disastrous. The predicted variation in the gravitational constant also was compatible with existing observations. Because the universe evolves on a timescale of 10 to 20 billion years, then if G is determined by the distribution of matter in the universe through the scalar field, it would be expected to vary at approximately the same rate, or at about 1 part in 20 billion per year. The predicted effect of such a variation on the structure of the Sun, Earth, or stars, or on the orbits of planets or the Moon was small enough to be no problem, given the large observational errors that were usually present in attempts to measure such long-term, slowly changing phenomena.

Initially, the theory received a cool reception. Although some theorists were pleased with the way it incorporated Mach's principle, most saw no strong reason to overthrow general relativity. The real boost for the scalar-tensor theory came in 1966, with the Dicke-Goldenberg solar oblateness results. The inferred value for the flattening of the Sun and the resulting contribution to Mercury's perihelion advance of about 3 arcseconds per century fit naturally with the prediction of the scalar-tensor theory of 40 arcseconds, assuming a value of ω of 7. The total advance of 43 arcseconds then agreed with the observed value. Here for the first time was apparent evidence against general relativity and in favor of a serious alternative, and despite the controversy over the proper interpretation of the oblateness data, interest in the scalar-tensor theory, or the Brans-Dicke theory as it came to be called, began to grow.

Some theorists began to take the theory seriously and to apply it to astrophysical problems, such as the structure of neutron stars and black holes, the nature of gravitational radiation, and cosmological models, in the same manner as was being done with general relativity. The output of papers devoted to the Brans-Dicke theory grew. The total number of papers published on the subject between 1962 (the year of the first three papers by Brans and Dicke) and 1967, approximately five, was matched in each of the years 1968 through 1971, and between 1972 and 1975, the annual output was two and a half times again as many. Other theorists began to think about the question of testing general relativity in a new light. Previously they took general relativity for granted, but now they realized that they had to take a broader, more unbiased viewpoint, which might perhaps lead both to new experimental ideas and to new insights into general relativity itself. One indication of this new viewpoint was the joke that used to go around Kip Thorne's relativity research group at Caltech: On Monday, Wednesday, and Friday, we believe general relativity; on Tuesday, Thursday, and Saturday, we believe the Brans-Dicke theory (on Sunday, we go to the beach). The late 1960s and early 1970s represented the peak of interest and activity in the scalar-tensor theory. But during this period were sown the seeds of its demise.

For example, it was this unbiased, theory-independent viewpoint that allowed a theorist like Kenneth Nordtvedt to discover that a broad class of theories including the Brans-Dicke theory predicted a breakdown in the equivalence principle for massive self-gravitating bodies, an effect that ultimately was tested in favor of general relativity. In addition, the fact that the predictions of the theory differed from those of general relativity by only percents motivated more precise experiments than might otherwise have occurred. It was no longer adequate simply to detect an effect predicted by general relativity; one had to measure it with a high degree of precision. For instance, the existence of the theory played an important role in strengthening the scientific objectives of the tracking program for Mariner 6 and 7. Planning of such aspects of the mission as the number of radar observations near superior conjunction, the optimal data-analysis procedure, even the overall length of the mission, was influenced by the requirement to measure the time delay to better than the 10 percent precision that had been achieved to date by the passive radar measurements to Mercury and Venus. In order to distinguish between general relativity and the Brans-Dicke theory, 1 or 2 percent precision was needed. These and many other examples demonstrate the powerful impact that the scalar-tensor theory had on the subject of experimental relativity.

Results of these new experiments began to roll in during the early 1970s, and the fortune of the theory began to decline. By 1972, radio-wave light deflection experiments were agreeing with general relativity to 3 percent and better, forcing ω to be larger than 10, and by 1975 the agreement was to 1.5 percent. In 1975, Mariner 6 and 7 time-delay results were published, confirming general relativity to 3 percent. In 1978 came Mariner 9 results at 2 percent, and finally, in 1979 the best Viking results, at 0.1 percent. The latter placed a lower limit on ω of 500. Meanwhile, the lunar laser-ranging results were published in 1976, demonstrating that the Nordtvedt effect did not occur, in agreement with general relativity. The accuracy of the measurement corresponded to a lower limit on ω of 29. By the end of the decade, there was very little further interest in the theory as a serious alternative to general relativity, although some relativists continued to explore some of its theoretical aspects. After 1980, the annual output of papers on the theory leveled off at about eight, which represented a declining fraction of the total number of papers devoted to general relativity and its applications, which grew steadily. The problem of Mercury's perihelion shift and the solar oblateness remained unresolved; if anything it was now even more contentious, because the prediction of the Brans-Dicke theory with ω larger than 500 for Mercury's perihelion shift is indistinguishable from that of general relativity, so if the solar oblateness were to be as large as the original Dicke-Goldenberg 1966 value, both theories would be in violation of experiment.

Could one say that the scalar-tensor theory was completely dead? Not exactly. Because ω is adjustible, the predictions of the theory can be made to be as close as desired to those of general relativity; so as long as experiments continue to agree with general relativity, they will always agree within experimental error with the scalar-tensor theory for a sufficiently large value of ω. At this point a certain subjectivity must enter the decision as to what is viable and what isn't. Generally speaking, physicists have been guided in such situations by a principle known as Occam's razor. Enunciated by the fourteenth-century philosopher William of Occam, although it is traceable back to Aristotle, this principle states *"pluritas non est ponenda sine necessitate,"* or nature likes things as simple as possible. Actually, it is a known fact that physicists like things as simple as possible, and it is assumed that physicists are a reflection of nature.

This principle of simplicity still has a subjective side, for what one means by simple depends on one's point of view or frame of reference. Suppose for example we were to use Occam's razor to decide between general relativity and Newtonian theory (assuming falsely that they both agreed with experiment). From one point of view, Newtonian theory would appear to be much simpler. It involves only one gravitational potential and a few simple equations, whereas general relativity has the full machinery of curved space-time, with a metric tensor that describes space-time curvature having ten potentials instead of one. Occam's razor would presumably favor Newton. But consider a different point of view. Suppose we accept the fact, based on what we have learned from the Eötvös experiment, that space-time is curved, and, like it or not, we must use the mathematics of curved space-time in discussing gravitational theories. In this language, general relativity is actually rather simple. Its theoretical content is characterized by the space-time metric tensor, and a set of Einstein equations, and that's it. By contrast, Newtonian theory in this language turns out to be horribly complicated: the space-time metric tensor can only be partially defined, and two additional fields (one of them more complicated even than the metric tensor) must be introduced to make everything come out right. From the point of view of the devotee of curved space-time, Occam's razor would clearly favor general relativity. Of course, in this case, experiment alone is capable of deciding between Newtonian theory and general relativity.

In the case of the Brans-Dicke theory versus general relativity, both are curved space-time theories with similar structure. General relativity has only the metric tensor as its basic field, whereas the Brans-Dicke theory has both the metric tensor and the scalar field. Thus Occam's razor would favor general relativity. This doesn't necessarily mean that a scalar-tensor theory (with a large enough ω to agree with experiments) can never be viable; it simply means that general relativity provides a simpler description of nature in accord with observations, and should be used until there is some overriding reason to consider an alternative.

The story of the Brans-Dicke theory is a classic illustration of the scientific method at work, and the role of Robert H. Dicke is a classic example of the scientist's search for truth.

9

Is the Gravitational Constant Constant?

WHY shouldn't it be? After all, the gravitational constant G is a fundamental constant of nature. It is simply the proportionality constant that determines the gravitational force between two bodies of given masses and separation. This is the case, at least in Newtonian gravitation, but even in general relativity, G plays an analogous role as a constant of proportionality that determines the amount of space-time curvature produced by a given density of matter. Is there any reason to consider the possibility that a constant appearing in such a fundamental part of physics as the law of gravitation might actually vary with time?

Before 1929, the answer would be a definite no. But after 1929, you couldn't be so sure. The event that weakened our certitude that the gravitational constant was constant was the announcement that the universe was expanding.

Before this time, as far as anyone could tell, the universe at large was static and unchanging, at least over long time scales. For example, the laws of motion proposed by Aristotle around 300 B.C. and elaborated five hundred years later by Ptolemy had the stars stationary on a fixed celestial sphere with the Sun and planets orbiting the Earth on fixed circles called deferents. Some of the complicated wanderings of the planets were accounted for by having them move on smaller circles called epicycles, whose centers moved around the Earth on the deferents. Nevertheless, even though there were changes in the positions of the Sun and planets, the laws governing them were immutable. Even when the regime of Aristotle and Ptolemy was overthrown in the sixteenth century by the Sun-centered hypothesis of Nicholas Copernicus and by Kepler's laws of planetary motion, the laws themselves were again immutable. Finally, in Newton's law of universal gravitation, the quantity that we now call G was a fixed, universal constant. Because the universe far beyond the planets was known to be unchanging, there was no reason to think that the laws shouldn't be as well. General relativity also stated that the constant of gravitation was a true constant.

However, the discovery of the expansion of the universe generated a variety of suggestions that G might vary. To see how you might come up with such a suggestion, consider the Machian view of nature in which the local inertial or gravitational properties of matter are not absolute but are somehow connected or related to the distribution of matter at great distances. According to this viewpoint, the gravitational force between two bodies might depend on distant matter, so that if the universe is expanding and its average matter density is decreasing, then the strength of the force between the bodies might also vary with time. As we saw in the previous chapter, Dicke had exactly this in mind in the late 1950s when he began thinking about a varying G. But he was by no means the first. Speculations of this kind began almost immediately after the 1929 announcement by the astronomer Edwin Hubble that the galaxies were receding from us and each other at rates that were proportional to their distances.

Because the idea of a varying G is closely linked to the expansion of the universe, let me digress for a moment to describe general relativity's record in cosmology, the study of the universe as a whole. Until the 1960s this record was decidedly mixed, and this in itself may have been partly responsible for the persistence of the idea of a varying G.

Over the years since 1917, when Einstein first calculated a cosmological model using general relativity, the theory has bounced from failure to success, back to failure, and finally, in the middle 1960s, to success. The first cosmological failure of general relativity was in a sense a failure of will for Einstein. In 1917, he applied general relativity to the problem of cosmology. This in itself was a bold step, because in 1917, it was not yet known whether there was anything but a void outside our own Milky Way galaxy. For example, the Andromeda galaxy, then called a nebula, was still believed to lie inside the Milky Way. Yet Einstein assumed that the universe could be idealized as a homogeneous distribution of matter, with a density that was the same everywhere. Eventually, this assumption was seen to be valid, at least as a first approximation. For a variety of philosophical reasons, he chose a model for the universe that was closed, in the sense that any observer setting off on a straight line (a geodesic) would ultimately return to his starting place, as a consequence of the curvature of space. The model was finite, yet unbounded, in the same sense that a two-dimensional analogue, the surface of a balloon, is finite in area, yet has no boundaries within the surface. The final assumption he made was reasonable enough: The model should be static, unchanging in time. This certainly fit the observational situation in 1917.

But to his horror, he discovered that the theory did not admit any such solutions. The only solutions allowed were either expanding or contracting. In order to get the needed static solutions, he had to modify the original equations of the theory by adding a term called the cosmological term. With the modified equations, he found that he could obtain static models for the universe.

He later referred to this as "the biggest blunder" of his scientific life, for Hubble's discovery showed that the universe was indeed expanding, so that the cosmological term was unnecessary. In 1931, Einstein recommended that it be dropped, and that the original field equations be restored. In this roundabout way, the expanding universe was converted from a failure to a success for the theory. It is amusing to speculate what course general relativity and cosmology might have taken if Einstein had stuck with the theory as he had developed it, predicted in 1917 that the universe must be evolving with time, and then sat back to await confirmation by observations.

With the cosmological term banished, general relativistic cosmology gave a decent accounting of the expansion of the universe. However, by the late 1940s, the theory appeared to be in jeopardy again. If you trace the expansion backward in time, you reach an era when the universe had to be extremely dense and extremely hot. If you continue even farther back, you reach a stage where everything goes crazy, where variables such as the density of matter become infinite. This is the stage called the big bang, and represents the beginning of the universe as we know it. If you know the rate at which any two galaxies are moving apart and the distance between them, you can estimate how long ago they must have been at the same location. This gives an estimate of the time since the big bang, or the age of the universe. Now, the rate of expansion that astronomers had established by the 1940s implied an age of the universe of 2 billion years. There was only one problem. By examining radioactive elements in rocks, geologists had determined that the age of the Earth was at least 3.5 billion years, older than the universe itself.

This was an embarrassment, and it was partly responsible for the rise and popularity during the 1950s of the steady-state theory of the universe devised in England by the astronomers Fred Hoyle, Hermann Bondi, and Thomas Gold. The steady-state theory got around the embarrassment by postulating that the universe has existed forever (and is therefore quite a bit older than the Earth), and then squared the expansion with the idea of a steady state by postulating the continuous creation of matter in the void between the existing matter.

General relativity started to recover from this failure in the late 1950s. Astronomers began to find serious errors in the methods that had been employed to determine the distances to the galaxies used to study the expansion of the universe. The new distance determinations put these galaxies much farther away than had previously been thought. The effect of this was to increase the time the galaxies would have been expanding since the big bang. The age of the universe implied by these new measurements was then in a much more comfortable range of larger than 10 billion years. More recent determinations are between 15 and 20 billion years.

The real cosmological successes for general relativity came in the 1960s. First was the discovery of the cosmic fireball radiation in 1965 by Penzias and Wilson. This radiation appears to be the remains of the hot electromagnetic radiation that would have dominated the universe in its earlier phase, now cooled to 3° above absolute zero by the subsequent expansion of the universe. Second came calculations by theorists of the amount of helium that would be produced by the thermonuclear fusion of hydrogen in the very early universe, around 1,000 seconds after the big bang. The amounts, approximately 25 percent by weight, were in agreement with the abundances of helium observed in stars and in interstellar space. This was an important confirmation of the hot big bang picture, because the amount of helium believed to be produced by fusion in the interiors of stars was woefully inadequate to explain the observed abundances. These two results together with other observations spelled the death knell of the steady-state theory. Because the universe in that model is in a steady state, it was always as cold as it is today, while to generate the background radiation and the helium, a hot phase of the universe seems to be necessary. Today, the general relativistic hot big bang model of the universe has broad acceptance, and cosmologists now focus their attention on more detailed issues, such as how galaxies and other large-scale structures formed out of the hot primordial soup, and on what the universe might have been like earlier than 1,000 seconds, all the way to a trillion-trillion-trillionth of a second (and some brave cosmologists are going back even farther) when the laws of elementary particle physics played a major role in the evolution of the universe. Contrary to Einstein's original preference for a closed universe, the observational evidence currently favors an open, infinite universe that will expand indefinitely.

But back to G. Unlike the steady-state theory, which appeared to succumb relatively gracefully, the idea of a variable G, once raised, has lingered. As we will see, the idea is still around because it has proven very difficult to rule out experimentally.

One of the first to suggest that G should change was the British physicist Paul A. M. Dirac (1902–84). In addition to being one of the pioneers of the quantum theory of matter, Dirac also worked in general relativity, and speculated widely on a variety of subjects, from subatomic particles to cosmology. One of his speculations concerned what has come to be known as the large numbers hypothesis. Generally speaking, it is possible to take various constants of nature, each with its own dimensions or units, and combine them in various ways to produce numbers in which the units disappear; the numbers are then called dimensionless. One example might be the ratio of the mass of a proton to the mass of an electron. This number is approximately 2,000, and is dimensionless because it is a mass divided by a mass. No matter what units are used to express the individual masses (grams, pounds, stone), the ratio is always the same. This is what makes such numbers useful. Another example is the square of the unit of electric charge divided by the product of the speed of light with Planck's constant, the constant that appears in quantum mechanical descriptions of atomic and subatomic physics. This number is approximately 1/137. Most such dimensionless numbers are not all that far from unity (at least compared to the numbers I am about to quote).

However there are a few numbers of this sort that are very far from unity. For instance, consider the ratio of the electric force between an electron and a proton to the gravitational force between the two. Because both forces vary as the inverse square of the distance between the two particles, when you take the ratio, the distance cancels out, and the result is a number that depends only on G and the two masses (which appear in the gravitational force) and on the electronic charge (which appears in the electric force). The resulting number is enormous, approximately 10 to the 40th power (10^{40}). This is a result of the fact that the electromagnetic interaction is much, much stronger than the gravitational interaction, at least on the microscopic scale of electrons and protons. The only reason that gravity dominates on planetary and astronomical scales is that electromagnetisim has associated positive and negative charges that like to pair off, effectively negating the electric and magnetic fields that might be exerted at long range. On the other hand, as far as any experiment has shown, the masses that produce gravity come with only one sign (no antigravity), so the bigger the mass, the larger the gravitational forces.

Another dimensionless number would be given by the age of the universe, approximately 20 billion years, expressed as a multiple of some fundamental unit of time. The year, month, and day are not fundamental units of time, because they depend on the particular characteristics of the orbits and rotation states of the Earth and Moon. The second is not a fundamental unit because it was defined originally as a certain fraction of a day (1/86,400). The most fundamental unit of time would be one associated with atomic processes, because it would depend only on basic natural constants, such as the electric charge, the mass of the electron, or the speed of light. This time unit, which appears throughout physics as the basic time scale for atomic and nuclear processes, is roughly the time required for light to travel a characteristic distance sometimes called the classical electron radius (not the true radius of the electron). Its value is approximately 10^{-23} seconds. Thus the age of the universe in atomic time units is also approximately 10^{40}.

This is quite a coincidence. Or is it?

Of all the possible powers of 10 between 0 and 100, say, why should two such dimensionless numbers, arrived at by very different reasoning, come so close to each other? This suggested to Dirac that more than coincidence was at work here. He postulated that the near equality of these two numbers was a manifestation of some as yet unknown deeper law of nature that required them to be nearly equal for all time. There was one problem with this, however. The age of the universe is not constant; it is continuously increasing. Therefore, if the equality between the two large numbers quoted above is to be maintained always, then at least one of the other numbers—electron or proton mass, electronic charge, speed of light, or G—must change with time in an appropriate way. Very few physicists would accept the proposition that the electronic charge or mass, or the speed of light, or other such atomic constants were changing with time, and in fact there is a large amount of experimental evidence that these constants are indeed constant. The only plausible candidate to vary is G. If G varied with time in the right way, then the two large numbers could in principle maintain their equality for all time. Dirac never uncovered any deeper law of nature that demanded this equality, but once he made this large numbers hypothesis, it developed a life of its own, and has fascinated physicists ever since.

There is one interesting and unusual viewpoint on the large numbers hypothesis that does not require G to vary, however. It states essentially that the two large numbers quoted above are so close to each other, not because of coincidence, not because of a deeper law of nature that requires them to be equal always, but because we exist. Because this viewpoint makes the existence of human observers an integral part of the argument, it is sometimes called the anthropic principle. What made the large numbers hypothesis so interesting was that it connected a quantity that varies, the age of the universe, with quantities that we previously assumed were constant, such as atomic constants and G. However, there is an additional property of the age of the universe that is important, and that is that we exist now in order to observe that the universe is as old as it is. This replaces the question, Is the equality between the large numbers a coincidence? with the question, Is our existence now a pure coincidence? According to the anthropic principle, the answer to the latter question is no; we exist now because the values of G and the atomic constants are what they are.

The argument goes like this. The lifetimes of stars like our Sun are determined by the pull of gravity that squeezes the stellar material and heats it up, and by the atomic processes that determine how rapidly the heat and light are transported out of the stellar interiors and emitted to space. These lifetimes therefore depend both on the value of G and on the values of atomic constants such as the masses of proton and electron, and on atomic time scales, the very quantities that were ingredients in the large numbers quoted above. The lifetimes turn out to be on the order of billions of years. We know that stars produce their light by converting hydrogen and helium into heavier elements by the process of nuclear fusion, and we also know that such heavy elements as carbon, nitrogen, and oxygen are needed to produce astronomers. The only way we know of to get these heavy elements out of stars and into interstellar space where they can be condensed into planets and astronomers is during a supernova, the cataclysmic dying phase of a massive star, in which the outer layers are blasted into space, while the inner core collapses inward to form a neutron star or a black hole, objects that we will encounter again in later chapters. Therefore, in order to have observers capable of determining the age of the universe, we need to have had supernovas, and therefore the universe must be at least as old as a typical star, in other words, billions of years old. Thus, when the universe was 1,000 times younger than it is now (in other words only a few million years old), the two large numbers were not close to being equal because the one involving the age was 1,000 times smaller, but then we weren't around to measure them! Similarly, when the universe is 1,000 times older than it is now, the equality will again not be true, but by then all the stars will have died, leaving behind cold white dwarfs, neutron stars, and black holes, and all observers will have long since died, and again no one will be around to measure the numbers. Therefore, the near equality between the two 10^{40} numbers is a consequence of our being around now to measure the present age of the universe and the atomic constants.

Therefore, according to the anthropic principle, the gravitational constant G does not have to vary with time in order to understand the large numbers equality. However, we should be careful not to take the anthropic principle too seriously. The reason is that it is really not a physical theory: there is no way to test it. For example, we can't start the universe all over again with different values of the fundamental constants to see what happens. In this sense, the anthropic principle has no predictive power; it can only "postdict," take what we know has already happened and explain it in some manner, and even then in only a loose and qualitative way. You might say that the entire subject of cosmology suffers from the same defect of "postdiction." The crucial difference is that the cosmological endeavor attempts to be quantitative. For example, the standard big bang model of general relativity makes a quantitative statement about the amount of helium produced in the early universe: it is close to 25 percent, not 45 percent, and not 5 percent. It also makes other predictions (or postdictions if you prefer) that can be tested by further observation: for example, that the abundance of deuterium (the heavy isotope of hydrogen) is a very sensitive indicator of the average density in the universe both at the time of the synthesis of helium and today. If the observed abundance of deuterium in stars and in interstellar space (taking account of the deuterium that may have been created or destroyed by processes occurring since the big bang) is not in accord with the observed average density of matter in the universe, then the model may have to be modified or dropped. On the other hand, the very loose arguments and statements associated with the anthropic principle (such as, "astronomers exist") are not subject to such strict tests. For this reason, the anthropic principle has not played an important role in gravitation or cosmology. This does not mean to imply, however, that one cannot have a lot of fun with such anthropic speculations. (Question: Why is the universe isotropic, apparently the same in all directions? Anthropic answer: Of all cosmological models in general relativity, nonisotropic as well as isotropic, only the isotropic ones allow galaxies to condense properly out of the primordial matter, thereby allowing astronomers to exist to measure the isotropy of the universe.)

Dirac's large numbers hypothesis was not the only proposal leading to a variable G. There have been many over the years, dating back to the late 1920s. I have already mentioned the Machian approach taken by Dicke, in which the gravitational constant is assumed to be somehow related to the distribution of distant matter, and therefore changes with time as the density of matter decreases with the universal expansion. A similar idea was put forward in the early 1970s by Fred Hoyle and his Indian colleague Jayant Narlikar, in a new theory of gravity that was an outgowth of Hoyle's earlier steady-state cosmology.

Whether you are a devotee of the large numbers hypothesis, or a convert to the Brans-Dicke or Hoyle-Narlikar theories, you will find a common testable prediction: G varies with time. General relativity's prediction: G is constant.

But suppose it were to vary. Because the possible variation of G is tied to the evolution of the universe, we would naïvely expect G to vary at a rate corresponding to the rate of expansion or the rate of aging of the universe. Because the universe ages at a rate of one year per year, it therefore ages at a rate of about 1 part in 20 billion of its total age, per year. Thus, we might expect G to change by 1 part in 20 billion per year. As it turns out, most models with a varying G say that G should decrease with time, in other words, gravity should be getting weaker.

How would such a variation in G show up? If gravity were getting weaker, then stars and planets, which are held in by gravity, would expand. The Earth, for example, would then slow down in its rotation, just as the spinning ice skater slows down upon extending her arms away from her body, and the

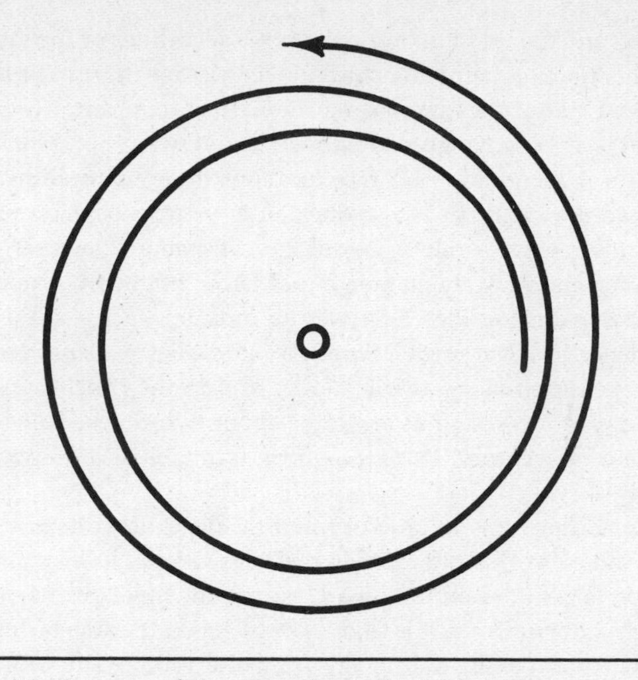

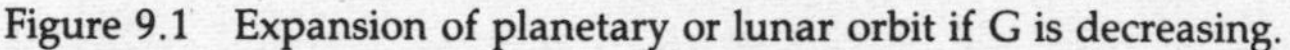

Figure 9.1 Expansion of planetary or lunar orbit if G is decreasing.

day would become longer. Similar conclusions would apply to the orbits of the planets and the Moon. As gravity became weaker the orbits would expand, and the planetary years and the lunar month would become longer (see figure 9.1). Now, before you begin to worry about how you are ever going to deal with a 107-hour day, remember that the rates of these processes, if they occur at all, are parts in 20 billion per year. During the entire age of the Earth (4.5 billion years), the length of the day might have increased by only 20 percent.

During the early 1960s, Dicke and his students carried out systematic studies of all the data that was known about such effects and concluded that a variable G at the above rate could not be ruled out, although there was no strong evidence in support of it, either.

One way to check for a variable G involves the motion of the

Moon and the rotation of the Earth. Observations of the motion of the Moon, whether by studying its passage in front of background stars, or by measuring the Earth-Moon distance using lunar laser ranging, give a relatively clear cut answer: The lunar month is increasing at a rate of about three-hundredths of a second per century. This corresponds to an expansion of the Moon's orbit at a rate of about 2.6 centimeters per year. You will remember from chapter 7 that Dicke originally supported the laser-ranging idea as a way to look for such an effect as evidence for a varying G. Similarly, studies of the length of the day measured using atomic clocks lead to the conclusion that the day is increasing at a rate of about two-thousandths of a second per century. Does this mean that G is decreasing? Not necessarily.

The difficulty is in how to interpret these observations, and the source of this difficulty is the tides. We all know that the Moon raises tides on the ocean because the pull it exerts on the Earth is not uniform; it is slightly stronger on the side facing the Moon than on the side away from the Moon. The Sun also raises tides that are a little less than half as large. Less well known because it is not as visible as the ebb and flow of the ocean at the beach is the fact that the Moon and Sun cause similar distortions of the Earth as a whole. These "solid Earth tides" are much smaller than the oceanic tides because the solid part of the Earth is more rigid than the oceans are. Because of this we can ignore them and assume that the solid part of the Earth is unchanged.

An important consequence of these tides is friction that causes the Earth to slow down its rotation, and the Moon to recede from the Earth. This idea was first put forward by the philosopher Immanuel Kant in 1754. Kant argued that tides must have been responsible for changing the Moon's original rotation rate to a rate equal to its rate of revolution around the Earth, so that it always presents the same face to us, and therefore, tides must ultimately have the same effect on the Earth.

To see how this comes about, imagine the following simple

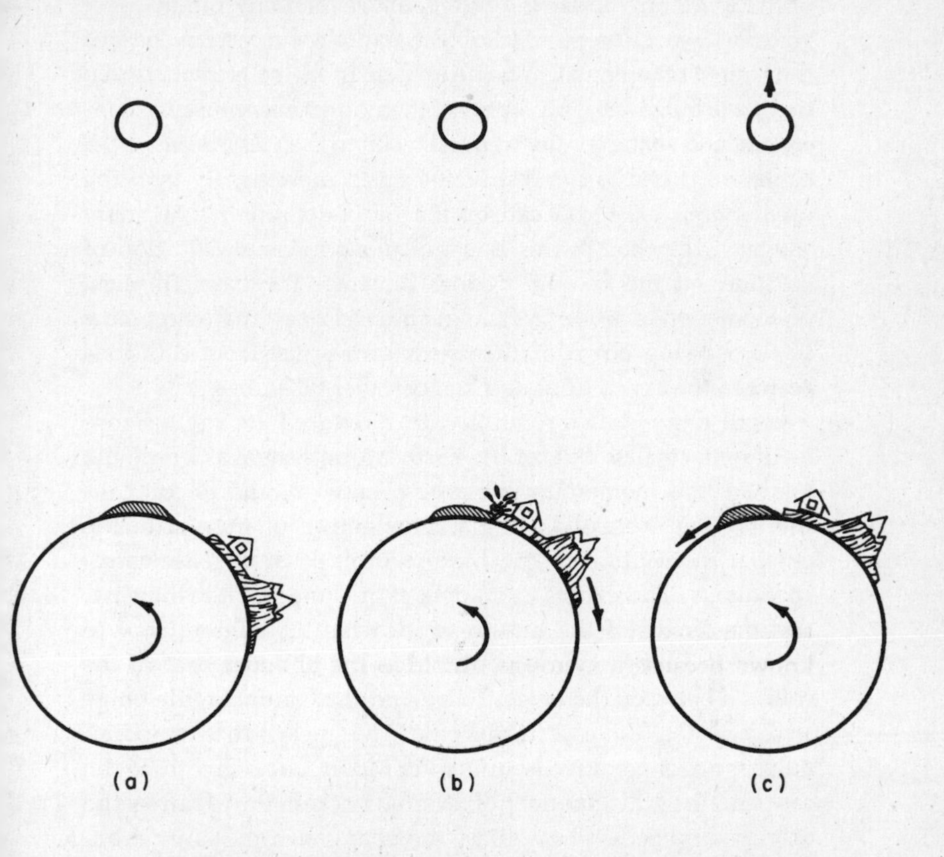

Figure 9.2 Tidal friction. (a) Moon creates a tidal bulge in the ocean that tries to stay directly underneath. (b) Rotation of the Earth causes collision between ocean bulge and land, causing the land to recoil and the Earth to slow its rotation slightly. (c) Ocean bulge also recoils and is no longer directly under the Moon; its gravitational force on the Moon causes the Moon to recede slightly.

picture (see figure 9.2). Suppose the Moon is above a portion of the North Atlantic ocean, and raises a tidal bulge in the water. To make the bulge, of course, water must flow into it from other parts of the ocean. The bulge wants to remain directly under the

Moon, because of the Moon's gravitational pull. However, the Earth is rotating under the bulge, and eventually the bulge of water will encounter a solid object, namely the eastern coast of the United States and Central America. In the collision between the tidal bulge and the continent, momentum must be conserved, and so the continent must recoil to the west. Because the continent is tied to the rest of the Earth more tightly than the oceans are, this recoil causes the Earth to rotate a bit more slowly. Even though this is a ridiculous oversimplification of the tides of the oceans, it does illustrate the basic physical phenomenon at work. In fact, geophysicists estimate that most of the slowing down of the Earth comes just from collisions between the ocean tides and the continental shores.

Because the Earth's rotation rate is reduced, its angular momentum is smaller. But for the Earth-Moon system as a unit, the total angular momentum must be conserved, and to compensate for the decreased angular momentum of the Earth, the angular momentum of the Moon's orbit must increase correspondingly. This causes the radius of the lunar orbit to increase, and the length of the month to increase. A different way to understand the response of the Moon is to notice that in the collision between the ocean bulge and the continent, the bulge also recoils, to the east in this case. This means that the ocean bulge is no longer directly under the Moon, but is slightly to the east, and the gravitational pull exerted back on the Moon by the displaced bulge is what causes the changes in the lunar orbit. All this seems eminently reasonable, and the numbers turn out to be in agreement with the observations, within errors of something like 25 percent.

But reasonable isn't good enough if we want to see whether G is changing at a rate of 1 part in 20 billion per year. Such a varying G would cause the lunar month to increase by at most two-hundredths of a second per century, or two-thirds of the observed amount. Its effect on the rotation rate of the Earth would be smaller because its partial rigidity will keep it from expanding freely as G decreases. Of course, if G varied at a slower rate, these effects would be even smaller. So we need somehow to be able to calculate the tidal effects with much better accuracy to see if there is a discrepancy with observations that could be attributed to a varying G. In the case of the rotation of the Earth, this goal is made even more difficult by a host of other possible effects that could cause the Earth's rate to change, such as variations in wind patterns and speeds, redistribution of the molten material in the Earth's core, and contraction of the Earth as it ages and cools.

These difficulties did not stop Thomas A. Van Flandern. Engaged in the study of the motion of the Moon and planets at the U.S. Naval Observatory, he believed he had a method of separating cleanly the effects of the tides from the effects of a changing G, in order to see if there was any of the latter. The origin of the idea is uncertain, although its roots date back to work in the late 1940s of Gerald M. Clemence, the great celestial mechanician of the U.S. Naval Observatory. It was most clearly expounded in 1972 by Fred Hoyle, as a way to test the variable-G prediction of the Hoyle-Narlikar theory.

The idea is deceptively simple. Assume for the moment that there are no tidal effects in the Earth-Moon system and that G is decreasing. As a consequence of the decreasing G, the lunar month will increase by a certain percentage (corresponding to 1 part in 20 billion) each year. However, because the orbital periods of the Earth and planets depend on G in exactly the same manner as does the orbital period of the Moon, then the periods or "years" of all those planets will increase by exactly the same percentage in the same amount of time. I should point out here that we are assuming that all these periods and times are being measured by atomic clocks, whose rates are fixed, independent of G. But if the month and year increase by the same percentage annually, then if we compare the length of the month directly with the length of the year, the two will not change relative to each other, even though each will increase relative to an atomic clock. This is easy to understand. Consider two poor physics graduate students with identical cheap Mickey Mouse watches that each run slowly by one minute per hour, relative to the fine electronic watch of their professor. The two students will always agree with each other, while steadily falling behind the professor.

Now let's introduce tidal friction into the Earth-Moon system. As it turns out, there are no significant tidal friction effects on the planetary orbits around the Sun. The length of the year will therefore increase on a percentage basis relative to atomic clocks only by the varying G effect. With tidal friction and a decreasing G, the length of the month will increase by the sum of the two effects, tidal and varying G. Therefore, when we compare the lunar month directly with the length of the year, only the tidal increase will be seen, because, as I remarked above, a pure decrease in G causes no relative change in these two times. All we then have to do is subtract the third value (month versus year, containing only tidal effects) from the second (month versus atomic time, containing tidal and varying G effects), watch the tidal piece cancel out, and presto, whatever is left over is due to a varying G.

Unfortunately, like most of the other gravitational experiments that I have described in this book, what looks simple in principle turns out to be terribly difficult in practice. The rate of change of the lunar month relative to the length of the year can be obtained from analysis of telescopic observations of the Moon and the planets dating back three centuries, and from records of ancient solar eclipses, dating back three thousand years. Because these observations involve comparing the position of the Moon with the positions of the Earth and other planets in their orbits, they give us the rate of increase of the month measured directly against the year, a number that should contain only the tidal contribution. Unfortunately, because these kinds of observations involve astronomical records dating back many centuries, they are subject to many sources of error, and numerous corrections must be applied to them, with the result that different experts have come up with different values for the rate of increase, ranging from 3.0 hundredths of a second per century to 5.8 hundredths of a second. The second set of data needed, the rate of increase of the month measured against atomic clocks, was not available before 1955, because atomic clocks did not exist before then. Once atomic clocks became available, astronomers at such places as the U.S. Naval Observatory and the Royal Greenwich Observatory in England could make these measurements. But here again, because the motion of the Moon is so complex and the analysis of the data difficult, the result was a range of values. At different times during the ten years that he worked on this problem, Van Flandern himself came up with a variety of values, such as 7.1 hundredths of a second per century in 1970, 8.9 hundredths in 1975, and 5.2 hundredths in 1976. Any difference between the latter set of numbers (month versus atomic clocks) and the former set (month versus year) would imply a changing G, remember, but we see that the spread in the values within any one set is as large as the possible differences between the sets, so it is difficult to make any strong conclusion one way or the other, despite the fact that the second set of numbers was generally somewhat larger than the first. For a time in the middle 1970s, Van Flandern claimed that his data indicated a real difference between the two values, with the Moon versus atomic time value somewhat larger than the Moon versus year value, and interpreted this to mean that G was decreasing at a rate of about 1 part in 25 billion per year. But because of the problems that I have described, his claim was eventually given little credence.

If tides are the main culprit complicating attempts to measure a variation in G using the Moon, then why not forget the Earth-Moon system altogether, and study just the motions of the planets? Their orbits are not subject to the tidal effects that plague the lunar orbit. If G is decreasing with time, then the orbits should expand, and the orbital periods should increase. Again, we have to measure these orbital periods against atomic clocks, not against the orbital period or year of the Earth, for the reasons that I outlined previously. This confines us again to the post-1955 period, which is all right because we also need very precise determinations of the orbits, and that requires radar ranging, which as we have seen, became available only after 1960.

Now, because the effect builds up orbit by orbit, the best planets to study are clearly those with the shortest periods. This means that we should concentrate on Mercury, Venus, and Mars. Early results, based primarily on radar-bounce observations of Mercury carried out by Shapiro's group at the Haystack and the Millstone Hill facilities between 1966 and 1969 showed no evidence for a changing G, down to the level 8 parts in 20 billion per year. Observations of Mercury and Venus up to 1974 improved this limit by a factor of 4.

But the best limit to date comes from radar ranging to Mariner 9 and the Viking spacecraft on and around Mars, the same method that yielded such precise measurements of the Shapiro time delay. Two circumstances made this improvement possible; first, the improved accuracy in the Earth-Mars distance, down to the 10-meter level, and second, the completely fortuitous and unexpected longevity of the Viking Lander *1*. Designed as a nine-month device for the purpose of TV photography and biological experiments, the spacecraft simply refused to die. From 1976 until July 1982, when its batteries became so weak that NASA controllers reluctantly gave the command to turn it off, this hardy beacon steadfastly transponded back to Earth the range signals sent to it. Without this six-year span of data, sensitive to the possible build-up of the effect of a varying G, the improved measurement would have been impossible. The result, obtained in separate analyses by the JPL group and Shapiro's group, was again no evidence for a variation in G, down to a level 10 times smaller than the previous determination, or to 1 part in 100 billion per year.

Actually, the description of these planetary range measurements given above hardly does justice to them. As was the case in measuring the time delay, not only must the actual range measurements (over one thousand from the Viking landers) be taken, analyzed, and stored, but also the perturbations of the other planets on the orbit of the planet under study must be taken into account. Just as these perturbations produce perihelion shifts in the orbit of Mercury, as we saw in chapter 5, and just as they alter the position of the planet or spacecraft being used to measure the Shapiro time delay, as we saw in chapter 6, so here do they cause changes in the diameter and period of the orbit of the planet in question that look just like the effect of a changing G. (They had the same effect on the lunar orbit and also had to be taken into account in the analyses of variations of the lunar month.) Therefore, a tremendous amount of painstaking analysis, primarily using the huge computer codes devoted to analyzing the orbits of all the planets, must be done to eliminate these effects and to look for any residual effect of a varying G. It turns out that one of the main factors limiting the accuracy of the Viking results was the perturbing effect of the asteroids, the belt of interplanetary residue that lies primarily between the orbits of Mars and Jupiter. The largest of the asteroids, Ceres has a diameter one-third that of the Moon, while the other three thousand or so that have been catalogued range down in size from that. For all anyone knows, there may be 50 times that many altogether. To determine the perturbations they produce, we need to know their masses, but except for the few largest asteroids, this is a matter of educated guesswork. Thus, even though their effects on the orbit of Mars are tiny, they are significant, and this uncertainty is enough to limit the accuracy of the Viking determination.

Unlike the claims of Van Flandern, these results consistently showed no significant variation in G, to the accuracies quoted. Have we answered the question, Is the gravitational constant constant? No, we are really just on the verge of doing so. Remember our naïve expectation based, for example, on the large numbers hypothesis was that if G varied, it would do so at the rough rate of 1 part in 20 billion per year. The limit from the Viking observations was a part in 100 billion per year, only a factor 5 smaller than our naïve guess. We could easily imagine doing a detailed calculation within some specific theory such as the Brans-Dicke theory, and after keeping track of all the factors of 2 and π, coming up with a predicted rate of less than 1 part in 100 billion, thereby predicting a variable G that is still consistent with the observations. In other words, the Viking limit is not that much smaller than our guess to make it truly interesting.

However, to place a limit another factor of 10 smaller than Viking, or 50 times smaller than our guess, would be interesting. Such a limit, corresponding to less than 1 part in a trillion per year, would make a variable G much less plausible. Of course, as in the case of the Brans-Dicke theory, a variable G can never be completely ruled out, because measurements always have a finite error. But at an upper limit of 1 part in a trillion per year, we would be safe in employing Occam's razor, slicing away the idea of a variable G until a strong reason arises for reintroducing it.

Unfortunately, there is no current or planned experiment capable of achieving such a limit. Some theorists have carried out computer simulations and feasibility studies that suggest that a two-year span of radio range measurements to an orbiter around Mercury, with a range accuracy of 30 centimeters, could set a limit on a variation of G at the 3 parts in 10 trillion level. This would be a very interesting limit. Unfortunately, NASA has no plans for such a mission at present, or for the near future. Perhaps some sustained lobby by relativists in cooperation with planetary scientists, for whom such a mission would also be of interest, can change this situation.

10

The Binary Pulsar: Gravity Waves Exist!

IT IS UNLIKELY that Joe Taylor and Russell Hulse will ever forget the summer of 1974. It started uneventfully enough. Taylor, a young professor at the University of Massachusetts at Amherst, had arranged for his graduate student Hulse to spend the summer at the Arecibo Radio Telescope in Puerto Rico looking for pulsars. They had put together a sophisticated observational technique that would allow them to scan a large portion of the sky using the radio telescope in such a way that it would be especially sensitive to signals from pulsars. At that time, over one hundred pulsars were known, so their main goal was to add new ones to that list, in the hope that, by sheer weight of numbers, they could learn more about this class of astronomical objects. But apart from the possible payoff at the end of the observations, the bulk of the summer would be spent in rather routine, repetitive observing runs and compilation of data that, as in many such astronomical search programs, would border on tedium.

But on July 2, good fortune struck.

On that day, almost by accident, Hulse discovered something that would catapult both Hulse and Taylor into the astronomical headlines, excite the astrophysics and relativity communities, and ultimately yield the first confirmation of one of the most interesting and important predictions of general relativity. At least as far as relativists are concerned, the discovery ranks almost up there with the discovery of pulsars themselves.

The discovery of pulsars was equally serendipitous. In late 1967, radio astronomers Antony Hewish and Jocelyn Bell at Cambridge University were attempting to study the phenomenon of scintillation of radio sources, a rapid variation or "twinkling" of the radio signal from these sources that is caused by clouds of electrons in the solar wind out in interplanetary space. These variations are typically random in nature and are weaker at night when the telescope is directed away from the Sun, but in the middle of the night of November 28, 1967, Bell, who at the time was one of Hewish's graduate students, recorded a sequence of unusually strong, surprisingly regular pulses in the signal. After a month of further observation, she and Hewish established that the source was outside the solar system, and that the signal was a rapid set of pulses, with a period of 1.3373011 seconds. As a standard of time measurement, these pulses were as good as any atomic clock that existed at the time. It was so unexpected to have a naturally occurring astrophysical source with such a regular period that, for a while, they entertained the thought that the signals were a beacon from an extraterrestrial civilization. They even denoted their source LGM, for little green men. Soon the Cambridge astronomers discovered three more of these sources, with periods ranging from .25 to 1.25 seconds, and other observatories followed with their own discoveries. The little green men theory was quickly dropped, and the objects were renamed pulsars because of the pulsed radio emission.

This discovery had a tremendous impact on the world of astronomy. The discovery paper for the first pulsar was published on February 24, 1968 in the British science journal *Nature,* and in the remaining ten months of that year, over one hundred scientific papers were published reporting either observations of pulsars or theoretical interpretations of the pulsar phenomenon. In 1974, Hewish was rewarded for the discovery with the Nobel Prize in Physics, along with Sir Martin Ryle, one of the pioneers of the British radio astronomy program. In some circles, controversy still lingers over the decision of the Swedish Academy not to include Ms. Bell in the award.

Within a few years of the discovery, there was general agreement among theorists and observers about the overall nature of pulsars, although many of the details are still not completely ironed out. Pulsars are simply cosmic lighthouses: rotating beacons of radiowaves (and in some cases of optical light, X rays, and gamma rays) whose signals intersect our line of sight once every rotation period. The underlying object that is doing the rotating is a neutron star, a highly condensed body, typically of about the same mass as the Sun, but compressed into a sphere of around 20 kilometers in diameter, 500 times smaller than a white dwarf of a comparable mass. Its density is therefore about 500 million metric tons per cubic centimeter, comparable to the density inside the atomic nucleus, and its composition is primarily neutrons, with a contamination of protons and an equal number of electrons. Because the neutron star is so dense, it behaves as the ultimate flywheel, its rotation rate kept constant by the inability of frictional forces to overcome its enormous inertia. Actually there are some residual frictional forces between the neutron star and the surrounding medium that do tend to slow it down, but an example of how small this effect can be is given by the original pulsar: its period of 1.3373 . . . seconds is observed to increase by only 42 nanoseconds per year. Of the one hundred or so pulsars known by 1974, every one obeyed the general rule that it emits radio pulses of short period (between fractions of a second and a few seconds), and with a period that is extremely stable, except for a very, very slow increase. We will see that this rule almost proved to be the downfall of Hulse and Taylor.

Why a neutron star? Was this just a figment of the theorist's imagination, or was there some natural reason to believe in such a thing? In fact, neutron stars did begin as a figment of the imagination of the astronomers Walter Baade and Fritz Zwicky in the middle 1930s, as a possible state of matter one step in compression more extreme than the white-dwarf state. Such highly compressed stars, they suggested, could be formed in the course of a supernova, a tremendous explosion of a star in its death throes, that has been known to occur from time to time in our galaxy. While the outer shell of such a star explodes, producing a blaze of light that can momentarily exceed the light output of the entire galaxy and ejecting a fireball of hot gas, the interior of the star implodes until it has been squeezed to nuclear densities, whereupon the implosion is halted, leaving a neutron star as the cinder of the supernova. The neutron star should also be spinning very rapidly, for the following reasons. Most stars for which decent data exist are known to rotate, the Sun being the nearest example. Therefore, just as the figure skater spins more quickly when she pulls in her arms, exploiting the conservation of angular momentum, so too should the collapsing, rotating core of the supernova.

Of the five supernovas in our galaxy of which we have historical records during the past thousand years, one occurred in the constellation Taurus in 1054. It was recorded by Chinese astronomers as a "guest star" that was so bright that it could be seen during the day. The remnant of the supernova is an expanding shell of hot gas known as the Crab nebula. The observed velocity of expansion of the gas is such that, if traced backward in time for about nine hundred years, it would have originated in a single point in space. Now, several months after the discovery of the first pulsars, radio astronomers at the National Radio Astronomy Observatory trained the telescope on the central region of the Crab nebula and detected radio pulses. The discovery was confirmed at Arecibo, and the pulse period was measured to be 0.033 seconds, the shortest period for a pulsar known at the time. Compared to other pulsars, the Crab pulsar was slowing down at an appreciable rate, around 10 microseconds in period per year. This rate can be viewed differently; the time required for the period to change by an amount comparable to the period itself is around 1,000 years, which is just the approximate age of the pulsar, if it was formed in the 1054 supernova. Furthermore, if the pulsar is a rotating neutron star, the friction between it and the surrounding medium required to slow down its rotation at the observed rate was just enough to keep the nebula of gas hot enough to glow with the observed intensity. The fact that all these observations were so consistent with one another provided a beautiful confirmation of the rotating neutron star model for pulsars.

Other aspects of pulsars are not so clean cut or so simple, however, and one of these is the actual mechanism for the "lighthouse beacon," if indeed that is how the radio pulses are produced. In the conventional model, a pulsar is thought to have one important feature in common with the Earth: its magnetic northern and southern poles do not point in the same direction as its rotation axis. In the Earth, for instance, the magnetic northern pole is near Hudson's Bay, Canada, not in the middle of the Arctic Ocean, as is the northern pole of the rotation axis. There is one key difference, however. The magnetic field of a pulsar is a trillion times as strong as that of the Earth. Such enormous magnetic fields produce forces that can strip electrons and ions from the surface of the neutron star and accelerate them to nearly the speed of light. This causes the particles to radiate copiously in radio waves and other parts of the electromagnetic spectrum, and because the magnetic field is strongest at the poles, the resulting radiation is beamed outward along the northern and southern magnetic poles. Because these poles are not aligned with the rotation axis, the two beams sweep the sky, and if one of them hits us, we call it a pulsar. The details of this mechanism are extremely difficult to work out, partly because we have absolutely no laboratory experience with magnetic fields of such strength and with bulk matter at such densities, so the calculations rely heavily on theory (and on the computer).

Nevertheless, despite the difficulties in developing a complete picture of pulsars, by the summer of 1974 there was consensus on their broad features. They were rapidly rotating neutron stars whose periods were very stable except for a very slow increase with time. It was also clear that the more pulsars we knew about and the more detailed observations we had, the better the chances of unraveling the details.

This is what motivated and guided Hulse and Taylor in their pulsar search. The receiver of the 1,000-foot radio telescope at Arecibo was driven so that as the Earth rotated, in 1 hour, the instrument could observe a strip of sky 10 arcminutes wide by 3° long. At the end of each day's observations, the recorded data were fed into a computer, which looked for pulsed signals with a well-defined period. If a candidate set of pulses was found, it had to be distinguished from terrestrial sources of spurious pulsed radio signals, such as radar transmitters and automobile ignition systems. The way to do this was to return later to the portion of the sky to which the telescope was pointing when the candidate signals were received and see if pulses of almost exactly the same period were present. If so, they had a good pulsar candidate that they could then study further, such as by measuring its pulse period to the microsecond accuracy characteristic of other pulsars. If not, forget it.

The day-to-day operation of the program was done by Hulse, while Taylor made periodic trips down from Amherst throughout the summer to see how things were going. On July 2, Hulse was by himself when the instruments recorded a very weak pulsed signal. If the signal had been more than 4 percent weaker, it would have fallen below the automatic cutoff that had been built into the search routine and would not even have been recorded. Despite its weakness, it was interesting because it had a surprisingly short period, only 0.059 seconds. Only the Crab pulsar had a shorter period. This made it worth a second look, but it was August 25 before Hulse got around to it.

The goal of the August 25 observing session was to try to refine the period of the pulses. If this were a pulsar, its period should be the same to at least six decimal places, or to better than a microsecond, over several days, because even if it were slowing down as quickly as the Crab was, the result would be a change only in the seventh decimal place. Then the troubles began. Between the beginning and the end of the two-hour observing run, the computer analyzing the data produced two different periods for the pulses, differing by almost 30 microseconds. Two days later, he tried again, with even worse results. As a result, he had to keep going back to the original discovery page in his lab notebook and cross out and re-enter new values for the period. Hulse's reaction was natural: annoyance. Because the signal was so weak, the pulses were not clean and sharp like those from other pulsars, and the computer must have had problems getting a fix on the pulses. Perhaps this source was not worth the hassle. If Hulse had actually adopted this attitude and dumped the candidate, he and Taylor would have been the astronomical goats of the decade. As it turned out, the suspicious Hulse decided to take an even closer look.

During the next several days, Hulse wrote a special computer program designed to get around any problems that the standard program might be having in resolving the pulses. But even with the new program, data taken on September 1 and 2 also showed a change in pulse period, a steady decrease of about 5 microseconds during the 2-hour runs. This was much smaller than before, but still larger than it should be, and it was a decrease instead of the expected increase. To continue to blame this on the instruments or the computer was tempting, but not very satisfying.

But then Hulse spotted something. There was a pattern in the changes of the pulse period! The sequence of decreasing pulse periods on September 2 appeared to be almost a repetition of the sequence of September 1, except it occurred 45 minutes earlier. Hulse was now convinced that the period change was real and not an artifact.

But what was it? Had he discovered some new class of object: a manic-depressive pulsar with periodic highs and lows? Or was there a more natural explanation for this bizarre behavior?

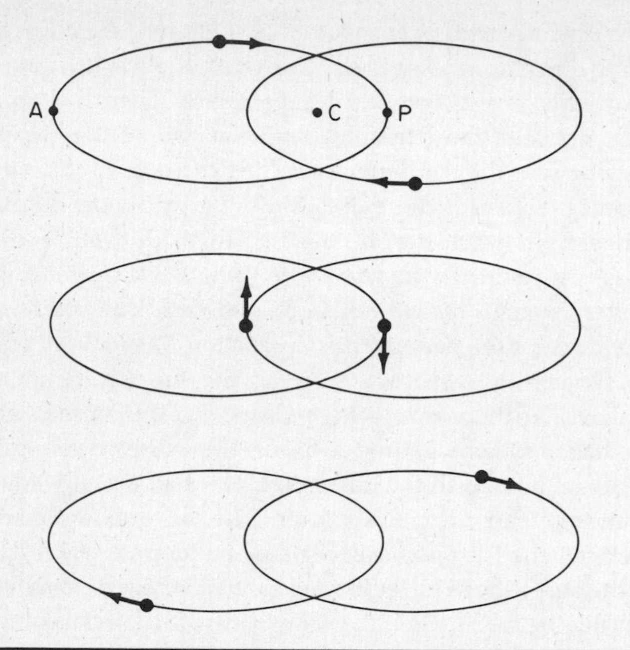

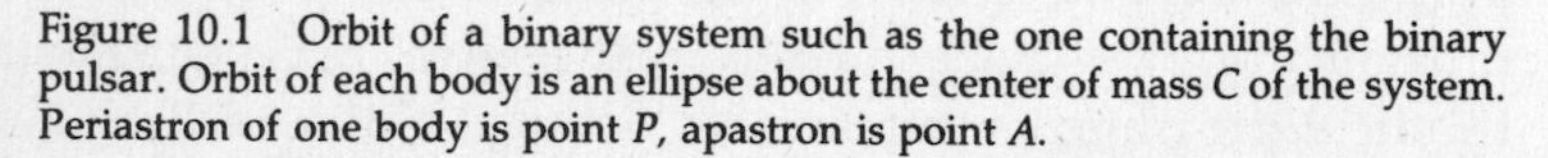
Figure 10.1 Orbit of a binary system such as the one containing the binary pulsar. Orbit of each body is an ellipse about the center of mass *C* of the system. Periastron of one body is point *P*, apastron is point *A*.

The fact that the periods nearly repeated themselves gave Hulse a clue to an explanation. The source was indeed a well-adjusted pulsar, but it wasn't alone!

The pulsar, Hulse postulated, was in orbit about a companion object, and the variation in the observed pulse period was simply a consequence of the Doppler shift (see figure 10.1). When the pulsar is approaching us, the observed pulse period is decreased (the pulses are jammed together a bit) and when it is receding from us the pulse period is increased. Actually, optical astronomers are very familiar with this phenomenon in ordinary stars. As many as half the stars in our galaxy are in binary systems (systems with two stars in orbit about each other), and because it is rarely possible to resolve the two stars telescopically, they are identified by the up-and-down Doppler shifts in the frequencies of the spectral lines of the stars. Here the pulse period plays the same role as the spectral line in an ordinary star. In most ordinary stellar binary systems, the Doppler shifts of the spectra of both stars are observed; however, occasionally one of the stars is too faint to be seen, so astronomers can detect the motion of only one of the stars. Such appeared to be the case here. One of Hulse's problems with this hypothesis was a practical one: he couldn't find any decent books on optical stellar binary systems in the Arecibo library because radio astronomers don't usually concern themselves with such things.

Now because the Arecibo telescope could only look at the source when it was within 1 hour on either side of the zenith or overhead direction (thus, the 2 hour runs), Hulse couldn't just track the source for hours on end; he could only observe it during the same 2-hour period each day. But the shifting of the sequence of periods in the September 1 and 2 data meant that the orbital period of the system must not be commensurate with 24 hours, and so each day he could examine a different part of the orbit, if indeed his postulate was right. On Thursday, September 12, he began a series of observations that he hoped would unravel the mystery (see figure 10.2).

On September 12, the pulse period stayed almost constant during the entire run. On September 14, the period started from the previous value and decreased by 20 microseconds over the 2 hours. The next day, September 15, the period started out a little lower and dropped 60 microseconds, and near the end of the run it was falling at the rate of 1 microsecond per minute. The speed of the pulsar along our line of sight must be varying, first slowly, then rapidly. The binary hypothesis was looking better and better, but Hulse wanted to wait for the smoking gun, the clinching piece of evidence. So far the periods had only decreased. But if the pulsar is in orbit, its motion must repeat itself, and therefore he would eventually be able to see a phase of the orbit when the pulse period increased, ultimately returning to its starting value to continue the cycle.

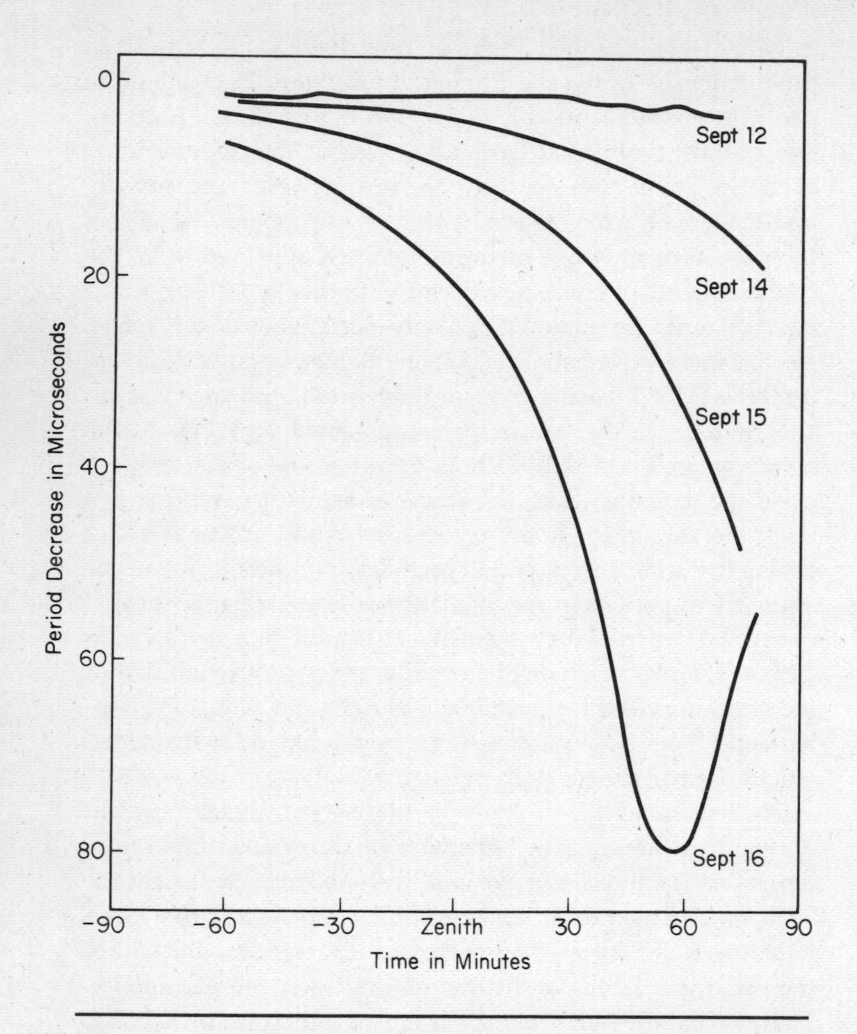

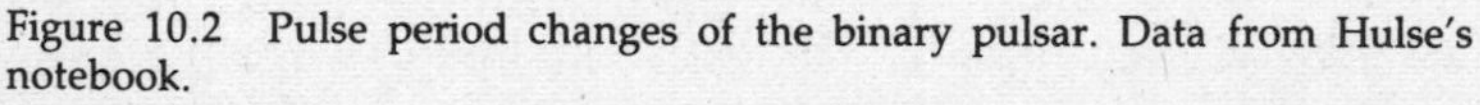
Figure 10.2 Pulse period changes of the binary pulsar. Data from Hulse's notebook.

He didn't have long to wait. The very next day, September 16, the period dropped rapidly by 70 microseconds, and with only about 25 minutes left in the observing run, it suddenly stopped decreasing, and within 20 minutes it had climbed back up by 25 microseconds. This was all Hulse needed, and he called Taylor in Amherst to break the news. Taylor flew immediately down to Arecibo, and together they tried to complete the solution of this mystery. However, the real excitement was still to come.

The first thing they determined was the orbital period, by finding the shortest interval over which the pattern of pulse readings repeated themselves. The answer was 7.75 hours, so the 45-minute daily shift that Hulse had seen was just the difference between three complete orbits and one Earth day.

The next obvious step was to track the pulse-period variations throughout the orbit to try to determine the velocity of the pulsar as a function of time. This is a standard approach in the study of ordinary binary systems, and a great deal of information can be obtained from it. If we adopt Newtonian gravitation theory for a moment, then we know that the orbit of the pulsar about the center of mass of the binary system (a point somewhere between the two, depending upon their relative masses) is an ellipse with the center of mass as the focus. The orbit of the companion is also an ellipse about this point, but because the companion is unseen, we don't need to consider its orbit directly. The orbit of the pulsar lies in a plane that can have any orientation in the sky. It could lie on the plane of the sky, in other words, perpendicular to our line of sight, or we could be looking at the orbit edge on, or its orientation could be somewhere between these extremes. We can eliminate the first case, because if it were true, then the pulsar would never approach us or recede from us and we would not detect any Doppler shifts of its period. We can also forget the second case, because if it were true, then at some point the companion would pass in front of the pulsar (an eclipse) and we would lose its signal for a moment. No such loss of the signal was seen anywhere during the 8-hour orbit. So the orbit must be tilted at some angle relative to the plane of the sky.

That is not all that can be learned from the behavior of the pulsar period. Remember that the Doppler shift tells us only the component of the pulsar velocity along our line of sight; it is unaffected by the component of the velocity transverse to our line of sight. Suppose for the sake of argument, that the orbit were a pure circle. Then the observed sequence of Doppler shifts would go something like this: Starting when the pulsar is moving transverse to the line of sight, we see no shift; one-quarter period later it is moving away from us, and we see a negative shift in the period; one-quarter cycle after that it is again moving transverse, and we see no shift; one-quarter cycle later it is moving toward us with the same velocity, so there is an equal positive shift in the period; after a complete orbital period of 8 hours, the pattern repeats itself. The pattern of Doppler shifts in this case is a nice symmetrical one, and totally unlike the actual pattern observed. The observed pattern tells us that the orbit is actually highly elliptical or eccentric. In an elliptical orbit, the pulsar does not move on a fixed circle at a constant distance from the companion; instead it approaches the companion to a minimum separation at a point called periastron (the analogue of perihelion for the planets) and separates from the companion one-half of an orbit later to a maximum distance at a point called apastron. At periastron, the velocity of the pulsar increases to a maximum and then decreases again, all over a short period of time, while at apastron, the velocity slowly decreases to a minimum value and slowly increases again. The actual behavior of the Doppler shift with time indicated a large eccentricity (see figure 10.3). Over a very short period of time (only 2 hours out of the 8) the Doppler shift went quickly from zero to a large value and back, while over the remaining 6 hours, it changed slowly from zero to a smaller value in the opposite sense and back. In fact, the September 16 smoking gun observation saw the pulsar pass through periastron, while the September 12 observations saw the pulsar moving slowly through apastron. Detailed study of this curve showed that the separation between the two bodies at apastron was 4 times larger than their separation at periastron. It also showed that the direction of the periastron was almost perpen-

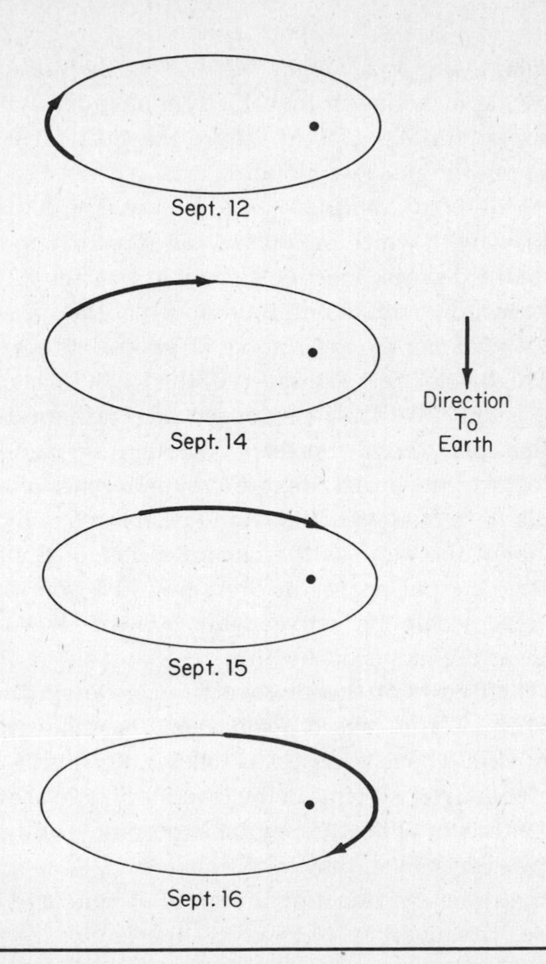

Figure 10.3 Location of pulsar in its orbit. On Sept. 12, the pulsar is moving through apastron; its speed is low and slowly varying, so there is little change in the observed period (see figure 10.2). The pulsar is moving away from us, so the period is longer than the "rest" period. On Sept. 14, the pulsar is moving almost transversely, so there is little Doppler shift, and the period is lower than before. On Sept. 15, the pulsar is starting to move toward us and is speeding up as it nears periastron; the pulse period decreases markedly toward the end of the run. On Sept. 16, the pulsar starts out moving almost transversely, then quickly passes through periastron, so its velocity toward us quickly reaches a maximum, then decreases; the pulse period rapidly reaches a minimum and then increases. The portion of the orbit seen during the same two-hour interval each day varies because the orbital period is 7.75 hours, so the portion seen is forty-five minutes further advanced each day.

dicular to our line of sight, because the periastron (the point of most rapid variation in velocity) coincided with the largest Doppler shift (the point where the pulsar has the smallest amount of transverse motion).

At this point, things began to get hot. The actual value of the velocity with which the pulsar was approaching us, as inferred from the decrease in its pulse period, was about 300 kilometers per second, or about one-thousandth of the speed of light! The velocity of recession was about 75 kilometers per second. These are high velocities! The speed of the Earth in its orbit about the Sun is only 30 kilometers per second. Furthermore, if 200 kilometers per second represents a rough average of the orbital velocity of the pulsar, then the circumference of an orbit that it could trace out in 8 hours would be about 6 million kilometers, or about the same as the circumference of the Sun. In other words, the pulsar was in orbit around a companion with an average separation between the two that was only about as large as the radius of the Sun.

When news of this discovery began to spread in late September 1974, it caused a sensation, especially among general relativists. The reasons are as follows. Relativists are always on the lookout for systems in the laboratory or in astronomy where the effects of relativity may be important. They determine this in two ways. First, they take a characteristic speed of an object in the system, divide it by the speed of light, and square it. The closer this quantity is to unity, the larger are the effects of special relativity, and the happier they are. For the Hulse-Taylor pulsar, using 200 kilometers per second, this quantity is about 5 parts in 10 million. This is not all that large, except when you compare it to the corresponding value for Mercury, an object where we know relativistic effects are important, for example in the perihelion shift. For Mercury (48 kilometers per second), this quantity is only 2.5 parts in 100 million, a factor of 20 smaller. Secondly, relativists like to take a characteristic mass of an object in a system, multiply it by the gravitational constant, divide by a characteristic size (in this case the separation between the bodies), and divide by the speed of light squared. This number is a rough measure of how large the deviations from flat space-time are near the system, and the bigger it is, the better. For a black hole, it is 0.5. Using two solar masses as the mass of two neutron stars, we find that this number for the pulsar binary system is around 4 parts in 1 million, while for Mercury, using the mass of the Sun and the orbital radius of Mercury, it is also 2.5 parts in 100 million. This made relativists happy indeed, for on the face of it, the orbit of this pulsar was 20 to 100 times more relativistic than Mercury's orbit. But that wasn't all. Because the orbital period of the pulsar was only 8 hours, more than 1,000 orbits would occur each year, so that any relativistic effect that built up orbit after orbit, such as the periastron shift (the binary system analogue of the perihelion shift), would build up over 250 times faster than an otherwise equivalent effect for Mercury, which makes only 4 orbits per year.

It was immediately clear that this new system, called the binary pulsar, was a new laboratory for observing general relativistic effects, and it was unique because it was the first such laboratory outside the solar system. During the fall of 1974 relativists and astrophysicists swamped the editorial offices of *The Astrophysical Journal Letters* with papers extolling the virtues of this new system and describing all the relativistic effects that could be observed in it. During an 8-week period in early 1975, this journal, designed for rapid publication of fast-breaking astronomical results, published seven papers of this kind, in addition to the Hulse-Taylor discovery paper. Between 1975 and 1977, over forty papers reporting either observational results or theoretical interpretations were published in a variety of astronomical journals, not quite the size of the output over the original pulsar, but still a significant cottage industry of research on one object.

Even before Hulse and Taylor's paper on the binary pulsar appeared in print (but too late to stop the presses), Taylor and his colleagues had detected the first of several important relativistic effects, the periastron shift of the orbit. As we have already seen, from the initial observations of these Doppler shifts of the pulsar period, it was clear that the periastron line, the line between the two bodies at minimum separation, was perpendicular to our line of sight: in other words, it lay in the plane of the sky. However, as time went on, the value of the maximum approach velocity began to decrease and the value of the maximum recession velocity began to increase: in other words, when the pulsar was heading purely toward us or purely away from us, its velocity was no longer the maximum or minimum allowed, but was somewhere in between in each case. That is, the periastron and apastron directions were no longer perpendicular to the line of sight, but had rotated slightly, by about one-third of a degree in one month. During a two and a half month observing program that ended on December 3, 1974, they tried to pin down this rotation. Coming up was the seventh installment of the Texas Symposium on Relativistic Astrophysics that had begun in Dallas in 1963. After cycling twice through a trio of cities that included Austin and New York, it was back in Dallas. The data analysis was completed just in time for Taylor to reveal to the audience on December 20 that the rate of periastron advance for the binary pulsar was 4.0° ± 1.5° per year. He would return to the Texas symposium four years later with an even more impressive announcement.

This periastron advance is about 36,000 times larger than the perihelion advance for Mercury, in keeping with what we expected: a factor 20 to 100 in the raw size of relativistic effects, and a factor of 250 in the number of orbits per year. Is this another triumph for general relativity? It is, but not in the obvious sense. The trouble is, the prediction of general relativity for the periastron advance for a binary system depends on the total mass of the two bodies; the larger the mass, the larger the effect. It also depends on other variables, such as the orbital period and the ellipticity of the orbit, but these are known from the observations. Unfortunately, we do not know the masses of the two bodies with any degree of accuracy. All we know is that they are probably comparable to that of the Sun in order to produce the observed orbital velocity, but there is enough ambiguity, particularly in the tilt of the orbit with respect to the plane of the sky, to make it impossible to pin the masses down any better from the Doppler shift measurements alone. Well if we can't test general relativity using the periastron shift measurement, what good is it?

It is actually of tremendous good, because we can turn the tables and use general relativity to weigh the system! If we assume that general relativity is correct, then the predicted periastron shift depends on only one unmeasured variable, the total mass of the two bodies. Therefore, the measured periastron shift tells us what the total mass must be in order for the two values to agree. From the fall 1974 observations, the inferred total mass was about 2.6 solar masses. Eventually, the periastron shift could be measured so accurately, 4.2263° per year, that the total mass of the system was pinned down to 2.8275 solar masses. This was the triumph for general relativity. Here, for the first time, the theory was used as an active tool in making an astrophysical measurement, in this case the determination of the mass of a system to a few parts in 10,000.

The relativists' intuition that this system would be a new laboratory for Einstein's theory was confirmed. But there was more to come.

During the first few months of observations of the pulsar, it was realized that this was a very unusual pulsar, over and above its being in a binary system. Once the periodic variations in its observed pulse period were seen to be due to Doppler shifts resulting from its orbital motion, these variations could be removed from the data, allowing the observers to examine the intrinsic pulsing of the object, as if it were at rest in space. Its intrinsic pulse period was 0.05903 seconds, but if it was slowing down as do other pulsars, it was doing so at an unbelievably low rate. It took almost an entire year of observation to detect any change whatsoever in the pulse period, and when the data were finally good enough to measure a change, it turned out to be only a quarter of a nanosecond per year. This was 50,000 times smaller than the rate at which the Crab pulsar's period changed. Clearly, any friction that the spinning neutron star was experiencing was very, very small. At this rate, the pulsar would change its period by only 4 percent in a million years. The steadiness and constancy of this pulsar made it one of the best timepieces the universe had ever seen!

This made it possible for the observers to measure the changes produced in the period by the orbital motion of the pulsar with better and better accuracy. The pulsar was so steady that Taylor and his colleagues could keep track of the radio pulses as they came into the telescope, and even when they had to interrupt the observations for long periods of time, as long as six months, while they returned to their home universities for such mundane duties as teaching, or while the telescope was used for other observing programs, they could return to the telescope after such breaks and pick up the incoming train of pulses, without losing track of a single beep. Eventually, the accuracies with which they could determine the characteristics of the pulsar and the orbit began to boggle the mind: for the intrinsic pulsar period, 0.059029995271 seconds; for the rate at which the intrinsic pulse period was increasing, 0.273 nanoseconds per year; for the rate of periastron advance, 4.2263° per year; for the orbital period, 27906.98163 seconds. Because the pulsar period changes by the quoted amount in the last three digits each year, the measured pulsar period is usually referred to a specific date, in this case, September 1, 1974.

There was more to this accuracy than just an impressive string of significant digits. This accuracy also yielded two further relativistic dividends. The first of these was another example of applied relativity, or relativity as the astrophysicist's friend. Beside the ordinary Doppler shift of the pulsar's period, there are two other phenomena that can affect it, both relativistic in nature. The first is the time dilation of special relativity: because the pulsar is moving around the companion with a high velocity, the pulse period measured by an observer foolish enough to sit on its surface (he would, of course, be crushed to nuclear density) is shorter than the period observed by us. In other words, from our point of view the pulsar clock slows down because of its velocity. Because the orbital velocity varies during the orbit, from a maximum at periastron to a minimum at apastron, the amount of slowing down will be variable, but will repeat itself each orbit. The second relativistic effect is the gravitational red shift, a consequence of the principle of equivalence, as we have already seen. The pulsar moves in the gravitational field of its companion, while we the observers are at a very great distance; thus, the period of the pulsar is red shifted, or lengthened, just as the period (or the inverse of the frequency) of a spectral line from the Sun is lengthened. This lengthening of the period is also variable because the distance between the pulsar and the companion varies from periastron to apastron, and it also repeats itself each orbit. The combined effect of these two phenomena is a periodic up and down variation in the observed pulsar period, over and above that produced by the ordinary Doppler shift. But whereas the Doppler shift changed the pulse period in the fifth decimal place, these effects, being relativistic, are much smaller, changing the pulse period only beginning at the eighth decimal place. It is extremely difficult to measure such a small periodic variation, given the inevitable noise and fluctuations in such sensitive data, but within four years of continual observation and improvement in the methods, the effect was found, and the size of the maximum variation was 58 nanoseconds in the pulse period. Again, as with the periastron, this observation does not test anything, because the predicted effect turns out to contain another unknown parameter, namely the relative masses of the two bodies in the system. The periastron shift gives us the total mass, but not the mass of each. Therefore we can once again be "applied relativists" and use the measured value of this new effect to determine the relative masses. The result is that the two masses must be very nearly equal, so that if the total mass is

2.8275 solar masses, the individual masses must be 1.42 solar masses for the pulsar, and 1.40 solar masses for the companion, good to about 2 percent. The use and understanding of relativistic effects here played a central role in the first precision determination of the mass of a neutron star.

These results for the masses of the two bodies were also interesting because they were consistent with what astrophysicists thought about the companion to the pulsar. Because it has never been seen directly, either in optical, radio, or X-ray emission, we must use some detective work to guess what it might be. It certainly cannot be an ordinary star like the Sun, because the orbital separation between the pulsar and the companion is only about a solar radius. If the companion were Sun-like, the pulsar would be plowing its way through the companion's outer atmosphere of hot gas, and this would cause severe distortions in the radio pulses that must propagate out of this gas, distortions that are not seen. Therefore, the companion must be much smaller, yet still have 1.5 times the mass of the Sun. Such astronomical objects are called "compact" objects, and astrophysicists know of only three kinds: white dwarfs, neutron stars, and black holes. The currently favored candidate for the companion is another neutron star, based on computer simulations of how this system might have formed from an earlier binary system of two massive stars that then undergo a series of supernova explosions that leave two neutron-star cinders. The fact that both masses turn out to be almost the same is consistent with the observation that in these computer models, the central core of the presupernova star tends to have a mass close to 1.4 solar masses. After the outer shell of each star is blown away, the leftover neutron stars each have about this mass. This mass is called the Chandrasekhar mass, after the astrophysicist Subrahmanyan Chandrasekhar, who determined in 1930 that this value was the maximum mass possible for a white dwarf (this discovery earned "Chandra" a share of the Nobel Prize in Physics in 1983). Because a presupernova core is similar in many respects to a white dwarf, it is not surprising that this special mass crops up here as well.

So why don't we see the companion? Because the binary pulsar is estimated to be about 16,000 light years away, neither a white-dwarf companion nor a blob of hot gas falling into a black hole would be bright enough to be detectable on Earth. A neutron-star companion would also be much too faint to be seen, unless it, too, were a pulsar. However, there is absolutely no evidence for any pulsed radio waves other than those from the main pulsar, so if the companion is a pulsar, its rotating beam must be pointing off in some other direction. Perhaps some distant advanced civilization with its own Hulse and Taylor is watching that pulsar and speculating on the nature of its companion!

But the biggest payoff of the binary pulsar was yet to come. To understand what this payoff was, we must turn back, first to 1916, then to the late 1960s, and finally to Munich, Germany, in 1978.

Einstein was not content simply to publish the general theory of relativity and to let matters end there. He continued for several years to study some of the consequences of the theory before turning most of his attention toward his ill-fated search for a unified field theory. One of these consequences was gravitational waves. Now, according to standard Newtonian gravitational theory, the gravitational interaction between two bodies is instantaneous, but according to special relativity, this should be impossible, because the speed of light represents the limiting speed for all interactions. Because general relativity was designed to be compatible with special relativity at some level, you would expect the theory to incorporate such a limiting speed for gravitational interactions. This means, for instance, that if two bodies are interacting gravitationally and one body suddenly changes its shape (from a sphere to a cigar, say), thereby producing a different gravitational force field, the effect of that change won't be felt instantly by the second body, but instead can only be felt within a region that expands outward from the first body at the speed of light. Eventually the change will be sensed by the second body. If the first body then returns to its original shape, the resulting change in the force field will also make its way outward at the speed of light. This phenomenon is the same as what happens when you jerk the end of a rope. The result is a wave that propagates along the rope at a speed given by such variables as the tension of the rope and its weight. In general relativity, the result is a gravitational wave, a wave of gravitational force that propagates outward at the speed of light. In chapter 12, we will talk about the nature of gravitational waves in a little more detail. To Einstein, the question was, Do the equations of general relativity indeed admit such a phenomenon, and if so, what would be the properties of such waves?

In fact, the equations did admit gravitational waves as solutions. For example, a dumbbell rotating about an axis passing at right angles through its handle will emit gravitational waves that travel at the speed of light. But Einstein also found that the waves have a very important property: They carry energy away from the rotating dumbbell, just as light waves carry energy away from a light source. He even derived a formula to determine the rate at which energy would be lost from a system, such as a rotating dumbbell, as a consequence of the emission of gravitational waves. As it turned out, the assumptions that he made to simplify the calculation were not completely valid, and he also made a trivial mathematical error that made his answer 2 times too large, but the basic analysis was correct. (The error was pointed out by Eddington.)

Einstein's paper on gravitational waves was published in 1916, and that was about all that was heard on the subject for over forty years. One reason was that the effects associated with gravitational waves were extremely tiny, just how tiny we will see in chapter 12. Another reason was that for a long time, there was disagreement over whether the waves were "real," or whether they were some artifact of the mathematics that would not have observable consequences. But by 1960, the beginning of the relativity revival, two developments resurrected the idea of gravitational radiation. One was the rigorous proof by relativity theorists that gravitational radiation was in fact a physically observable phenomenon, that gravitational waves do carry energy, and that a system that emits gravitational waves should lose energy as a result. The second was the decision by Joseph Weber of the University of Maryland to begin to build detectors for gravitational waves, not from dumbbells, but from extraterrestrial sources. But more about that later.

By 1974, gravitational radiation was a hot subject, and relativists were dying to find some. Even though Weber had claimed detection of waves as early as 1968, later experiments by other workers had failed to confirm his results, and the general feeling was that gravitational waves had not yet been found. Therefore, when the binary pulsar was discovered, and it was seen to be a new laboratory for relativistic effects, it seemed like a godsend. For if a rotating dumbbell can emit gravitational waves, then so can the rotating binary system, even though the two balls of the dumbbell are held together by a rod, whereas the two stars of the binary system are held together by gravity (in general relativity it doesn't matter what holds them together). The binary pulsar could be used in the search for gravitational waves.

But not in the obvious sense. Because the binary pulsar is 16,000 light years away, the gravitational radiation that it emits is so weak by the time it reaches the Earth that it is undetectable by any detectors of today or the forseeable future. On the other hand, if the waves are carrying energy away from the system, it must be losing energy. How will that loss manifest itself? The most important way it will manifest itself is in the orbital motion of the two bodies, because after all, it is the orbital motion that is responsible for the emission of the waves. A loss of orbital energy manifests itself in a speed-up of the two bodies and a decrease in their orbital separation. This seemingly contradictory statement can be understood when you realize that the orbital energy of a binary system has two parts: a kinetic energy associated with the motion of the bodies, and a gravitational potential energy associated with the gravitational force of attraction between them. So although a speed-up of the bodies causes their kinetic energy to increase, a decrease in separation causes their potential energy to decrease by about twice as much, so the net is a decrease in energy. The same phenomenon happens, for example, when an Earth satellite loses energy because of friction against the residual air in the upper atmosphere; as it falls toward Earth it goes faster and faster, yet its total energy is declining, being lost in this case to heat. In the case of the binary pulsar, the speeding up combined with the decreasing separation will cause the time required for a complete orbit, the orbital period, to decrease.

Here was a way to detect gravitational radiation, albeit somewhat indirectly, and a number of relativists pointed out this new possibility in the fall of 1974, soon after the discovery of the binary pulsar. As I mentioned previously, and as we will see in chapter 12, the effects of gravitational radiation are exceedingly weak, and this was no exception. The predicted rate at which the 27,000 second orbital period should decrease was only on the order of some tens of millionths of a second per year. Although this was an exciting possibility, the small size of the effect was daunting, and some thought it would take ten to fifteen years of continual observation to detect it. Perhaps by 1990. . . .

Now flash forward four years, to December 1978: the Ninth Texas Symposium on Relativistic Astrophysics, this time in Munich, Germany (Munich is in the province of Bavaria, sometimes considered the Texas of Germany). Joe Taylor was scheduled to give a talk on the binary pulsar. Rumor had it that he had a big announcement, and only a few insiders and theorists active in the subject of the binary pulsar knew what it was (I knew because I was scheduled to follow Taylor to present the theoretical interpretation of his results). A press conference had been set up for later in the day. The scene could have been London, November 6, 1919, where a similarly anticipated observational result bearing on Einstein's theory was about to be reported to the Royal Society. Taylor's announcement did not cause quite the same hoopla as the announcement of the light deflection measurements, yet it was no less important. It represented the climax of two decades of intensive testing of general relativity.

In a succinct, fifteen-minute talk (a longer, more detailed lecture was scheduled for the following day), Taylor presented the bottom line: after only four years of data taking and analysis, they had succeeded in detecting a decrease in the orbital period of the binary system, and the amount agreed with the prediction of general relativity, within the observational errors. This beautiful confirmation of an important prediction of the theory was a fitting way to open 1979, the centenary year of Einstein's birth.

It turned out that the incredible stability of the pulsar clock, together with some elegant and sophisticated techniques for taking and analyzing the data from the Arecibo telescope that Taylor and his team had developed, resulted in such improvements in accuracy that they were able to beat by a wide margin the projected timetable of ten years to see the effect. These improvements at the same time allowed them to measure the effects of the gravitational red shift and time dilation, and thereby measure the mass of the pulsar and of the companion separately. This was important because the prediction that general relativity makes for the energy loss rate depends on these masses, as well as on other known parameters of the system, so they needed to be known before a definite prediction could be made. With the values of about 1.4 solar masses for both stars, general relativity makes a prediction of 75 millionths of a second per year for the orbital period decrease. Using data taken through August 1983, Taylor and colleagues recently reported an observed value of 76 ± 2 millionths of a second per year.

As a young student of seventeen at the Polytechnical Institute of Zurich, Switzerland, Einstein studied closely the work of the nineteenth-century physicists Hermann Helmholtz, James Clerk Maxwell, and Heinrich Hertz, the pioneers of electromagnetism. Ultimately, his deep understanding of electromagnetic theory served him well in his attempts to formulate special and general relativity. It appears that he was especially impressed by an experimental result: Hertz's 1887 confirmation that light and electromagnetic waves are one and the same thing. The electromagnetic waves that Hertz studied were in the radio part of the spectrum, at 30 million cycles per second (30 megahertz). It is amusing to notice that our story of the decades of testing general relativity began with radio waves, the 440-megahertz waves bounced off Venus, and ended with radio waves, the pulsed signals from the binary pulsar, observed at the Arecibo operating frequency of 430 megahertz.

During the two decades that closed on the centenary of Einstein's birth, his theory was put on the firing line, confronted by experimenters determined to test it, attacked by theorists proposing alternatives to it, and grabbed by astronomers wanting to use it. The theory passed all its tests with flying colors, with the only remaining enigma being the oblateness of the Sun. But this does not end the story. The confrontation between general relativity and observation will proceed using new tools, in new arenas, across new frontiers. These new frontiers for observational relativity are the subject of the next chapter.

11

The Frontiers of Experimental Relativity

FRONTIERS in science are like the geographical frontiers of early America: temporary boundaries between the known and the unknown that we cross, sometimes cautiously, sometimes at full steam. What lies across the boundary we don't know; it could be exciting or it could be dull, there could be surprises or there could be nothing contrary to our expectations. As we proceed into the unknown region and learn something of what resides there, we establish a new frontier between what is now known and what is yet unknown. It is part of human nature to strive forever to push the frontier farther, whether in exploration of new lands, in athletic competition, or in scientific research.

In the late nineteenth century, the frontier of experimental gravitation was the anomalous perihelion shift of Mercury. More than seventy years passed following Le Verrier's discovery of the anomaly before it was crossed and successfully explored by general relativity. In 1960, the frontier was the completion of the program of testing Einstein's three famous predictions. This frontier was also successfully crossed using the latest technological advances in measuring devices, the space program, and new theoretical insights, and the newly explored region yielded many surprises—new experimental tests of general relativity and new versions of the old tests, all with accuracies undreamed of before.

We now find ourselves at a new frontier of experimental relativity. To get us across this frontier, the experimentalist will be the main guide. In order to refine existing tests of general relativity and to perform some important tests that have not been done to date, he will be called upon to push the limits of experimental technique as far as they can go (and sometimes beyond). The role of the theoretician may be somewhat less important at this frontier, because the effects to be measured or detected are known quantities, and theorists don't expect too many surprises such as new tests that haven't been discovered yet. That doesn't mean there won't be surprises, of course, for despite what they tell their experimentalist colleagues, theorists aren't perfect. If there are surprises, they will only add to the excitement of experimental gravitation.

I could try to give a compendium of all the different aspects of this new frontier of experimental relativity, but I won't for two reasons. First, because this chapter would then be as long as the rest of the book, and second, because the frontiers of physics can move quickly, and much of what I said here would soon be out of date or just plain wrong. Instead, I will give a brief sampler, a taste, of some of the issues that confront us at the frontier, if only to spark the reader's interest in watching for further developments.

Where is the frontier today? Of course, we are always interested in improving the accuracy of any existing experiment, in obtaining one more decimal place in a measured number associated with general relativistic effects. In principle, a violation of general relativity could turn up at any level of accuracy, and if it did that would be extremely interesting (and, no doubt, controversial). But physicists generally prefer to find new ways to make measurements or to try to measure new effects, rather than simply to repeat and refine the old experiments. One example that we have already mentioned in chapter 5, but that lies across the frontier because it may be well into the future, is Starprobe, a space mission near the Sun that could pin down the solar oblateness and clarify the interpretation of Mercury's perihelion shift.

Another idea that is currently under study is to try to measure the deflection of light by the Sun to much better accuracy than has been done to date, perhaps to the level of one-millionth of an arcsecond (a microarcsecond). Remember that the measured deflection for a grazing ray of light is around 2 arcseconds. At the microarcsecond level, general relativity predicts a small, higher-order correction to the deflection that would then be measurable. One concept to try to do this is under consideration, primarily at the Center for Astrophysics at Harvard, and has been given the acronym POINTS, for Precision Optical Interferometry in Space. POINTS would be an orbiting telescope designed to employ the technique of interferometry used in radio astronomy that I described in chapter 4, only in this case it would operate at the much shorter wavelengths of optical light. Another idea that is being tossed around is to take an advanced version of the hydrogen maser clock that was flown on a Scout D rocket to measure the gravitational red shift (chapter 3), and fly it near the Sun on a Starprobe-type mission, in order to detect the higher-order correction to the red shift predicted by general relativity.

These are just a few of the ideas that are being considered for the next generation of relativity experiments. However, there is one important experiment that actually has been in progress since 1960, but whose final result still lies beyond the frontier. It is called the gyroscope experiment.

The gyroscope experiment may go down in the history of physics as one of the most difficult and one of the longest experiments. If all goes as currently planned, this important experiment will have involved almost one-third of a century in designing and building the apparatus, while the actual measurement will not take much more than a year. The gyroscope experiment can be said to have been conceived by three naked men basking in the noonday California sun in the closing weeks of 1959. The three were all professors at Stanford University in Palo Alto. One of them was the eminent theoretical physicist Leonard I. Schiff, well known for his pioneering work in quantum theory and nuclear physics. In the late 1950s, however, he, like Dicke, had become interested in gravitation theory. The second professor was William M. Fairbank, an authority on low-temperature physics and superconductivity, who had just arrived at Stanford in September of 1959, lured there from Duke University in North Carolina. The third was Robert H. Cannon, also a recent acquisition of Stanford, an expert in aeronautics and astronautics from MIT.

But before we learn how these naked professors came to formulate this experiment, let us first answer the question, What does a gyroscope have to do with relativity?

When we think of a gyroscope we imagine something like a spinning flywheel. If the flywheel spins rapidly enough, its axis of rotation always points in the same direction, no matter how we rotate the platform or laboratory in which it sits, as long as the gyroscope is mounted on the platform using gimbals that allow it to turn freely with a minimum of friction. In other words, the axis of the gyro always points in a fixed direction relative to inertial space or to the distant stars. This, of course, is the basic principle behind the use of gyroscopes in navigation of ships, airplanes, missiles, and spacecraft. However, according to general relativity, a gyroscope moving about through curved space-time near a massive body such as the Earth will not necessarily point toward a fixed direction; instead, its axis of spin will change slightly, or precess. Two distinct general relativistic effects can cause such a precession.

The first of these is called the "geodetic effect," and is a consequence of curved space-time. Our everyday experience with gyroscopes tells us that as a gyroscope moves through space, its spin axis should maintain the same direction, in other words a direction parallel to its previous direction. However, in curved space-time, parallel in the local sense does not necessarily mean parallel in the global sense, and so upon completing a closed path, the gyroscope axis can actually end up pointing in a different direction than the one it started with.

A simple way to see how this can happen is to return to our old friends the Spherelanders. Because these people are only two-dimensional, they can't really construct the right kind of gyroscopes, but as an alternative, they can take a little ruler and slide it about on their sphere in a way that keeps it parallel to its previous direction (see figure 11.1). The ruler's edge then plays a role analogous to the spin axis of our gyroscope. To demonstrate what can happen, the Spherelanders consider the following closed route: from a point at 0° longitude on the equator, move east along the equator to 90° longitude, then go due north to the North Pole, make a 90° turn, and go due south back to the starting point on the equator. Suppose the Spherelanders start the ruler out parallel to the equator, pointing east. When they reach the first turn, the ruler will still point east, and when they head north, the ruler will now be perpendicular to the path. At the North Pole, they make a 90° left turn, but now the ruler's direction is to their rear, which is to the north as they head south. When they reach the equator once again, the ruler, which has been kept parallel to itself all the way, now points north, whereas it started out pointing east. The curvature of the two-dimensional surface of Sphereland accounts for this precession, and we understand it without much difficulty. The difference between this example and the geodetic effect on a moving gyroscope is that it is the curvature of space-time and not just space that is important.

The geodetic effect has been around since the early days of general relativity. The first to calculate the effect was Willem de Sitter, the Dutch theorist who had played a pivotal role in bringing general relativity to the attention of Eddington and the British physics community. In a paper published in the *Monthly*

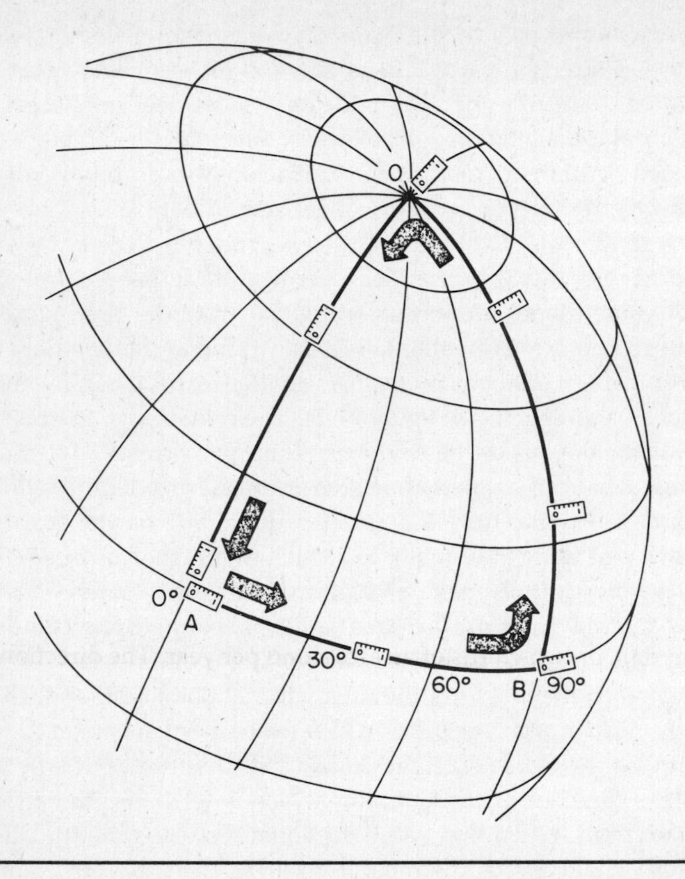

Figure 11.1 Precession of ruler in Sphereland. Along *AB*, ruler is carried parallel to itself from 0° longitude to 90° longitude. At *B* path turns north, but ruler continues to point east, and maintains direction up to North Pole. At *O*, a right turn of the path leaves ruler pointing to the rear. Ruler maintains direction to the rear back to *A*. Result is a precession of the direction of the ruler from easterly direction to a northerly direction.

Notices of the Royal Astronomical Society less than a year after Einstein's November 1915 papers on general relativity, de Sitter showed that relativistic effects would cause the axis of the orbit of the Earth-Moon system to precess at a rate of about 0.02 arcseconds per year. De Sitter was not thinking in terms of gyroscopes; instead, he had in mind how the combined relativistic gravitational fields of the Earth and Sun would perturb the Earth-Moon system. However, Eddington and others soon pointed out that the Earth-Moon system was really a kind of gyroscope, so the de Sitter effect was effectively a precession of the Earth-Moon gyroscope. Indeed, the Earth was also a gyroscope, so in fact both should precess in the same way. Unfortunately, the size of the effect was hopelessly small. Only today, with very high-precision radio interferometers that can measure the orientation of the Earth with respect to distant radio sources to the milliarcsecond level, can astronomers even begin to contemplate detecting such a small effect.

Instead of the Earth-Moon system, consider a more down-to-earth situation: a laboratory-size gyroscope at rest on the equator with its axis lying in the equatorial plane. As the Earth's rotation carries the gyroscope around through the Earth's curved space-time, the gyroscope will experience a precession in the plane at a rate of about two-thousandths of an arcsecond per day, or two-thirds of an arcsecond per year. The direction of the precession is in the same sense as the motion of the gyroscope around its path, counterclockwise if looking down from the North Pole. For a gyroscope that orbits the Earth in a low orbit, the amount of space-time curvature is not very different from what it is at the Earth's surface (see figure 11.2). But the orbiting gyroscope moves more quickly through the curved space-time than does the surface gyroscope, with a period of revolution of only about 1.5 hours, so the precession or change in direction will build up more quickly. Therefore, for a gyroscope in orbit at a few hundred miles altitude, the net precession in one year (5,000 orbits) will be around 6 arcseconds. This is the geodetic effect.

The other important relativistic effect on a gyroscope is known as the dragging of inertial frames, one of the most interesting and unusual of the predictions of general relativity (see figure 11.3). The origin of this effect is the rotation of the body in whose gravitational field the gyroscope resides. According to

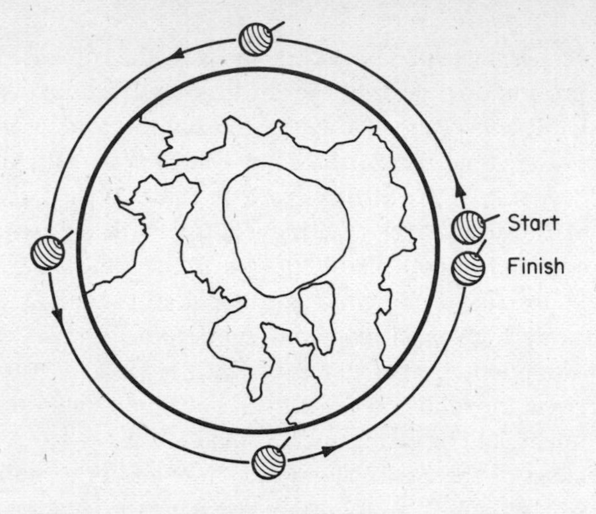

Figure 11.2 Geodetic precession of gyroscope in near-Earth orbit. After one orbit, direction of the gyroscope axis has rotated relative to its initial direction in the same sense (counterclockwise) as that of the orbit. Net effect over one year (5,000 orbits) is 6 arcseconds.

general relativity, a rotating body attempts to "drag" the space-time around it also in rotation. The simplest way to get a picture of the consequences of this dragging is to use a fluid analogy.

Consider a large swimming pool with a very large drain in its center. Water flows down this drain in a whirlpool of a kind we commonly see in bathtubs and sinks. To keep the level of the pool constant, let us assume that the water lost down the drain is continuously being replaced through inlets at the sides of the pool. Imagine now three Stanford professors floating in the pool. Professor Schiff is floating on an air mattress between the whirlpool and the edge of the pool, Professor Fairbank is floating on a similar air mattress but is straddling the whirlpool above the drain, and Professor Cannon is treading water. For simplicity's sake, let us also assume that each professor is anchored to the bottom of the pool by a tether attached to his waist. This is to prevent them from circling the drain, thereby

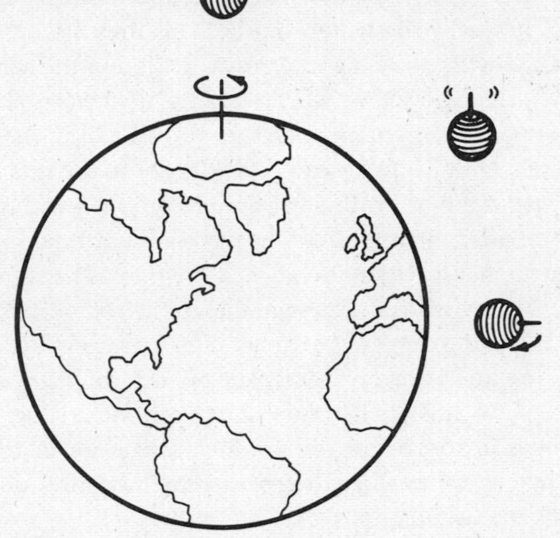

Figure 11.3 Dragging of inertial frames. Stationary gyroscopes near a rotating Earth can precess because of dragging of space-time by rotation of the Earth. If the axis lies perpendicular to the rotation axis of the Earth, the precession will be with the Earth's rotation for a gyroscope at the pole, and opposite to the Earth's rotation for a gyroscope at the equator. If the axis is parallel to the Earth's rotation axis, there is no precession. For other locations and other orientations of the gyroscope axis, the precession will be between these extremes.

complicating the effect we are looking for. With this arrangement, the behavior of these professors is very similar to that of three gyroscopes in a space-time being dragged by a rotating body. First consider Professor Schiff. Because the water closer to the whirlpool moves around more quickly than the water farther away, the foot of his air mattress is dragged more quickly than the head, and so while the whirlpool rotates, say counterclockwise as seen from above, Schiff's air mattress rotates or precesses in a clockwise direction. This is precisely the behavior of the axis of a gyroscope on the equatorial plane in a dragged space-time, with its axis pointing outward. Contrast this behavior with that of Professor Fairbank, whose air mattress straddles the whirlpool. The head and foot of his mattress are also pulled by the water, but because they are on opposite sides of the whirlpool, the mattress is pulled in the same sense as the whirlpool, in other words, counterclockwise. This is just what happens to a gyroscope on the rotation axis of the dragged space-time, with its own axis perpendicular to the rotation axis. Finally, we see that Professor Cannon, who is treading water, doesn't do much of anything. The direction of his body remains vertical, no matter where he goes in the pool. The same is true for a gyroscope whose axis is parallel to the rotation axis of the central body; the dragging of space-time has no effect on it. As with all the analogies for relativistic effects that I have used in this book, one must be careful not to push the analogy too far. Gyroscopes in space-time are not the same as air mattresses in water, but if the analogy helps us remember the qualitative effects, it is a useful one.

Of course, there is another key difference between the air mattress precession and the gyroscope precession due to the dragging of inertial frames. Size. The predicted precession for a gyroscope on the equator of the Earth is only one-tenth of an arcsecond per year. Unlike the geodetic precession, the dragging of inertial frames does not depend on whether or not the gyroscope is moving through space-time (the air mattresses precessed even though they were stationary in the pool), and so there is little difference in this case between the precession of a gyroscope on the Earth and that of a gyroscope in orbit. For a low Earth orbit, it is between 0.1 and 0.05 arcseconds per year, depending on the tilt of the orbit relative to the Earth's equatorial plane and on the initial direction of the gyroscope axis relative to the Earth's rotation axis.

What makes this effect so interesting and important is that while the other effects that I have described in this book, including the geodetic precession, have to do with such concepts as gravitational fields, curved space-time, and nonlinear gravity, this effect tells us something about the inertial properties of space-time.

If you ask yourself, "Am I rotating?," and you wish an answer with more accuracy than you can get simply by seeing if you are getting dizzy, you usually turn to a gyroscope, for the axis of a gyroscope is assumed to be nonrotating relative to inertial space. If you were to build a laboratory whose walls were constructed to be lined up with the axes of three gyroscopes arranged to be perpendicular to each other, you would conclude that your laboratory was truly inertial (and if it were in free fall, that would be even better). However, if your laboratory happened to be situated outside a rotating body, the gyroscopes would rotate relative to the distant stars because of the dragging effect I have just described. Therefore, your laboratory can be nonrotating relative to gyroscopes, yet rotate relative to the stars. In this way, general relativity rejects the idea of absolute rotation or absolute nonrotation, just as special relativity rejected the idea of an absolute state of rest.

To understand this more clearly, contrast it with Newtonian theory. True, Newtonian theory proposed that all inertial frames were equivalent, regardless of their state of motion, but it still had to allow an absolute concept when it came to rotation. As we saw in chapter 8 when I described Newton's bucket, the state in which the water is level in the bucket is, to the Newtonian physicist, a state of nonrotation relative to absolute inertial space, and the state in which the water climbs the side of the bucket is a state of rotation. Unfortunately, Newtonian theory found no way out of this absolutism when referring to rotation. Indeed, as of the early 1960s, it was not clear that general relativity provided a way out, and it was this conundrum that partly motivated Dicke to try an alternate approach to gravitation based on a scalar-tensor theory.

But in fact, as I suggested earlier, the dragging of inertial frames provides the way out of this absolutism. As early as 1923, Eddington suggested as much in his beautiful book on general relativity. However, it wasn't until the middle 1960s that theorists could show that the dragging effect provides an excellent accounting of how rotation is indeed relative.

The demonstration consisted of a simple model calculation of the following situation: Imagine a spherical shell of matter, like a balloon, that is rotating about some axis (for the purposes of this discussion we can ignore the flattening of the balloon caused by centrifugal forces). At the center of the shell is a gyroscope with its spin axis perpendicular to the axis of rotation of the balloon. According to Newtonian gravitation, the interior of the balloon is absolutely free of gravitational fields. The gyroscope feels no force whatsoever. To a first approximation, the same is true in general relativity, except for the dragging-of-inertial-frames effect, which produces forces in the interior of a rotating shell just as it would in the exterior. The effect of these forces is to cause the gyroscope to precess in the same direction as the rotation of the shell, but as you might imagine from our previous discussion, for a shell of planetary dimensions, say of the radius of a typical planet and containing a mass of a typical planet, the rate of precession is very small, much smaller than the rate of rotation of the shell. But now imagine increasing the mass of the shell and increasing its radius (keeping its rate of rotation the same), and consider the limit in which the mass tends toward the mass of the visible universe and the radius tends toward the radius of the visible universe. The remarkable result is that as you increase these values, the rate of precession of the gyroscope in the center grows, and in the limit, tends toward the rate of rotation of the shell. In other words, inside a rotating universe, the axes of gyroscopes rotate in step with the rotation; in other words, they are tied to the directions of distant bodies in that universe. Therefore, a laboratory tied to the gyroscopes, which we would define to be nonrotating, would indeed be nonrotating relative to the galaxies. If we had placed a bucket inside the shell instead of a gyroscope, and kept it fixed, or nonrotating, as we expanded our rotating shell, the frame-dragging forces would have caused the water to climb the sides of the bucket, so that an observer would see exactly the same physical phenomenon as would an observer looking at a rotating bucket inside a nonrotating universe. The existence of the dragging of inertial frames then guarantees that rotation must be defined relative to distant matter, not relative to some absolute space. This is what makes the detection of this effect so vital.

In this example, the interior of the shell was assumed to be empty except for the gyroscope, so the gyroscope's direction was tied rigidly to the distant galaxies of the shell "universe." If we now introduce a rotating body such as the Earth into the interior of the shell near the gyroscope, it will produce additional dragging-of-inertial-frames effects. Thus, the gyroscope will precess slightly relative to the distant galaxies because of the local dragging effect.

All well and good. But still the effects on gyroscopes on and near the Earth are horribly small. What would possess anyone to actually try to measure them? It is here that the three Stanford professors return to the story.

At Stanford University in the 1960s, back before the days of coeducational athletic facilities, the Encina gymnasium and its walled-in, open-air swimming pool was restricted to males only (the women's gym was on the other side of the campus). As such it was customary for users to swim in the nude. Schiff had a virtually unshakable routine of going to the Encina pool every day at noon, swimming 400 yards, and eating a bag lunch afterwards, while sunbathing. Even though he was chairman of the physics department, he would try his best to schedule meetings and appointments so as not to conflict with his noon-hour swim. Fairbank knew about Schiff's daily routine, and when he bumped into Cannon on campus one day in late 1959 and they began to talk about gyroscopes, Fairbank suggested that they go see Schiff at the swimming pool.

Each of these men had had gyroscopes on his mind for a while. Schiff had been thinking about gyroscopes ever since he opened his December 1959 issue of the professional physicists' magazine *Physics Today* and saw the advertisement on page 29. There, hovering in an artist's conception of interstellar space, was a perfect sphere, girdled by a coil of electrical wires, captioned "The Cryogenic Gyro." The advertisement announced the development at JPL of a super new gyroscope consisting of a superconducting sphere supported by a magnetic field (from the coils), all designed to operate at 4° above absolute zero. Schiff had taken a strong interest in tests of general relativity lately, and so he asked himself whether such a device could detect interesting relativistic effects. During the first two weeks of December he carried out the calculations, finding both the well-known geodetic effect, as well as the dragging-of-inertial-frames effect. The latter discovery was entirely new, at least as applied to gyroscopes. Back in 1918, two German theorists, J. Lense and Hans Thirring, had shown that the rotation of a central body such as the Sun would produce frame-dragging effects on planetary orbits that were unfortunately utterly unmeasurable, but no one had apparently looked at the effect of rotation of a central body on gyroscopes.

Fairbank's field was low-temperature physics, the properties of liquid helium, and the phenomenon of superconductivity, the disappearance of electrical resistance in many materials at low temperatures. He had also been thinking about the potential for a superconducting gyroscope that could be built in the new laboratory that he was setting up at Stanford, and he and Schiff had begun to talk about how these relativistic effects could be detected. Fairbank suggested measuring the precessions using gyroscopes in a laboratory on the equator, but this did not look promising. The reason was gravity. The best gyroscopes of the day had as their main element a spinning sphere, just as in the JPL advertisement. But the sphere had to be supported against the force of gravity, and the standard method of doing this was by electric fields or by air jets. Unfortunately, the forces required to offset gravity were so large that they introduced spurious forces or torques on the spinning ball that gave it a precession thousands of times larger than the effect being sought after, though easily small enough to permit accurate navigation and other commercial uses. This problem would effectively go away if the gyroscope were in orbit, where the gravitational forces are zero to high accuracy, and essentially no support is required. But remember, this was only two years after the launch of Sputnik, the first orbiting satellite, and Schiff and Fairbank could not imagine realistically being able to do this.

This was where Cannon came in. Cannon knew gyroscopes. He had helped develop gyroscopes used to navigate nuclear submarines under the Arctic icecap. He also knew aeronautics, and he was active in the fledgling space race that Sputnik had started. Before coming to Stanford from MIT, Cannon had already begun to consider the improvements in performance that would come with orbiting gyroscopes.

Finally the three were together (in the altogether) at the Stanford pool. When Schiff and Fairbank told Cannon about the proposed experiment, Cannon's first response was astonishment. To pull it off, they would need a gyroscope a million times better than anything that existed at that time. His next response was: Forget about doing it on Earth, put it into space! An orbiting laboratory is not at all farfetched, and in fact NASA was already laying plans for an orbiting astronomical observatory. Furthermore, Cannon knew the right people at NASA whom they could contact. With that, a three-decade adventure had begun.

It is another of those strange twists of scientific history that almost simultaneously with Schiff, Fairbank, and Cannon, someone else was thinking about gyroscopes and relativity. Completely independently of the Stanford group, George E. Pugh at the Pentagon was doing the same calculations. Pugh worked for a section of the Pentagon known as the Weapons Systems Evaluation Group, and for him, toying with gyroscopes was a perfectly reasonable activity, because gyroscopes have obvious military applications in the guidance of aircraft and missiles. In a remarkable memorandum dated November 12, 1959, Pugh outlined the nature of the two relativistic effects, although he had the frame-dragging effect wrong by a factor of 2, and described the requirements for detecting them using an orbiting satellite. Some of Pugh's ideas, such as a technique for compensating for atmospheric drag felt by the satellite, ultimately became important ingredients in the Stanford experiment. It is highly unlikely, however, that the Pentagon actually incorporated relativistic gyroscope effects into its military guidance systems.

In January 1961, Fairbank and Schiff kicked off the experiment officially with a proposal to NASA for an orbiting gyroscope experiment. The next two decades, until about 1983, were spent proving that it would work. The third decade, in which we now find ourselves, will involve building flight hardware. The actual experiment might happen by 1991. Sadly, Schiff's untimely death in 1971, at the age of 55, prevented his seeing the experiment reach fruition.

The goal of the experiment is to measure both the geodetic effect and the frame-dragging effect to an accuracy of better than a milliarcsecond per year. Because the smaller frame-dragging effect is only 50 milliarcseconds per year for the orbit being planned (a polar orbit), this means that a 2 percent measurement of this effect would be possible. The task of building an orbiting gyroscope laboratory that can measure such tiny effects has put the Stanford scientists at or beyond the frontiers of experimental physics and precision fabrication techniques, presenting them with apparently insuperable problems. Miraculously, they have managed to overcome each one.

A brief description of the current status of the experiment plan will illustrate the things that had to be done. The gyroscopes (for redundancy, there will be four) are spheres of fused quartz, about 4 centimeters in diameter. The spheres must be uniform in density and perfectly spherical in shape to better than 1 part in 10 million. This is like making the Earth perfectly spherical within a tolerance of plus or minus 3 feet. The reason for this requirement is that stray gravitational forces in the spacecraft, which are inevitably present even in orbit, will interact with any irregularities in the sphere and cause spurious precessions. Similar effects caused by tidal forces from the Sun and Moon acting on the Earth's equatorial bulge make the Earth's rotation axis precess. To overcome the problems of making a perfect sphere and then testing how spherical it is to the above precision required the invention of special new fabrication and testing procedures. If the balls are perfectly spherical, how do you determine the direction of their spin? You can't just attach a stick to the ball at one of the poles, because the stray gravitational forces acting on the mass of the stick would cause enormous precessions. The solution is to coat each ball with a thin, perfectly uniform layer of the element niobium. When the ball is spinning at low temperatures, near absolute zero, the niobium becomes a superconductor, its electrical resistance vanishes, and it develops a magnetic field whose north and south poles are exactly aligned with the spin axis of the ball. Very precise magnetometers, also operating near zero degrees absolute, can then determine the orientation of the magnetic field, and thereby of the spin axis. This required new techniques for working near absolute zero using liquid helium, and adapting those techniques to a space environment. Because the balls are perfectly spherical, how are they to be set spinning? The solution to this problem was to incorporate into the housing encasing each ball tiny jets that spray helium gas past the spheres, using friction to get them spinning. The helium gas comes from "boil off" from the liquid helium at 4° used to cool the apparatus. As I described previously, the gyroscopes will precess relative to the distant stars, so a very accurate telescope had to be designed and built into the spacecraft package to determine a reference direction to the milliarcsecond level per year. The current choice is the bright star Rigel in the constellation of Orion.

While simple to state in words, each of these problems was a major multi-year research and fabrication project. Overcoming each problem, and convincing NASA administrators and skeptical colleagues that the problems were surmounted are among the reasons it has taken so long to bring the experiment off. The experiment is now in an engineering development stage, which means that an apparatus is being built that can be tested and shaken down on a space shuttle flight planned for 1989. If that test is successful, an operational flight could take place as early as 1991.

The gyroscope experiment is perhaps an extreme illustration of the maxim that in experimental general relativity, nothing comes easy. But if Stanford and NASA can pull it off, the confirmation of the geodetic and frame-dragging precessions will have been worth the effort.

12

Astronomy after the Renaissance: Is General Relativity Useful?

THE TIME has come to take general relativity for granted. Not an unreasonable step, when you consider how the theory has successfully passed the various experimental tests that I have described in previous chapters.

The question is, Now that we believe it, what can we do with it? Naturally, one of the reasons for trying to test general relativity is that it is a fundamental theory of nature, and its validity has important epistemological and philosophical implications. Another reason is that in recent times general relativity has become a central player in the game of elementary particle physics, because the name of that game has become "unification": the unification of all the interactions of physics into one grand scheme, all compatible with quantum physics. These unification attempts do not alter the basic structure or predictions of general relativity in the situations that I have described in this book. It is only at super-high energies such as those imagined at the initial moment of the big bang that modifications of general relativity are anticipated. But these are subjects that I won't get into. Many excellent books discuss these topics; a few are listed at the end of this book as suggestions for further reading.

Instead, I want to ask what we can do with general relativity in the field that was one of the prime movers of the program to test the theory during the relativistic renaissance of the past twenty-five years: astronomy.

What we want to see is how general relativity can be used as a practical tool in understanding and unraveling some of the mysteries that astronomical observations reveal. We have already encountered some instances in which general relativity can be used this way. One was the case of gravitational lenses (chapter 4), where the general relativistic bending of light from a distant quasar could be used to infer something about the nature of the galaxy acting as the lens. Another instance was the binary pulsar (chapter 10), where we used general relativistic effects such as the periastron shift to help determine the masses of the two bodies in the system. There are many other ways in which applied general relativity can play a role in astronomy, but two of the most active and exciting at the moment are in the searches for black holes and gravitational waves.

Although the first glimmerings of the idea of the black hole can be seen, as we learned in chapter 4, in the eighteenth-century writings of Michell and Laplace, black-hole physics really didn't begin until 1939. That year, the physicist and future director of the U.S. atomic-bomb Manhattan Project, J. Robert Oppenheimer, and his co-worker Hartland Snyder, published a remarkable paper in *The Physical Review* entitled "On Continued Gravitational Contraction." The paper described what happens to a star that has exhausted the thermonuclear fuel necessary to produce the heat and pressure that support it against gravity. According to their calculations, the star begins to collapse, and if it is massive enough, it continues to do so until the radius of the star approaches a value called the gravitational radius, or Schwarzschild radius. This radius has a value given by twice the mass of the star times the gravitational constant G, divided by the square of the speed of light. For a body of 1 solar mass, the gravitational radius is about 3 kilometers; for a body of the mass of the Earth, it is about 9 millimeters. An observer sitting on the surface of the star sees the collapse continue to smaller radii, until both star and observer reach the single point at what was once the center of the star. On the other hand, an observer at great distances perceives the collapse to slow down as the radius approaches the gravitational radius, a consequence of the gravitational red shift of light and clocks that we encountered in chapter 3. However, in this case, the apparent slowing down becomes so extreme that the star appears almost to stop and hover just at the gravitational radius. The distant observer never sees any signals emitted by the falling observer once the latter is inside the gravitational radius. The calculations showed that any signal that the falling observer emits inside can never escape the sphere bounded by the gravitational radius.

The idea that there was something unusual about this gravitational radius did not originate with Oppenheimer and Snyder; it dates back almost to the inception of general relativity itself. Within two months of the publication of the final form of the theory, the German astronomer Karl Schwarzschild had obtained a pair of rigorous, exact solutions to the field equations, the first corresponding to an ideal body consisting of a mass point, the other corresponding to a spherical body of finite extent. The papers presenting these solutions were read before the Prussian Academy of Sciences by Einstein, because Schwarzschild was then in the German army at the Russian front. Tragically, Schwarzschild died there in May 1916 of an illness contracted at the front. In the first solution (conventionally called the Schwarzschild solution), the sphere at the gravitational radius appeared to be an unusual place because the mathematics of the solution went bad there (essentially some expressions in the space-time metric tensor became infinite).

Physicists often call such behavior "pathological." In the second solution, the gravitational radius lay inside the star, but the presence of the material making up the star altered the interior solution, so that the gravitational radius was a perfectly ordinary, nonpathological place. The second solution was an adequate solution for such bodies as the Sun, the Earth, in fact for any stationary, spherical body. The pathology of the first Schwarzschild solution led most relativists, Einstein included, to believe that it would be impossible for any body of a given mass to be so small in radius that the gravitational radius would lie outside it, thereby revealing this pathological surface to the external world. Despite the explicit demonstration by Oppenheimer and Snyder of a solution in which precisely such a thing would occur, their result was not taken particularly seriously, and the subject lay dormant for another two decades.

The revival of black-hole physics coincided with the renaissance of general relativity brought about in the 1960s, and was due to two things. The first, as I described in chapter 1, was the discovery of quasars. To understand the enormous energy output of these objects, theorists turned to the strong gravitational fields of superdense objects, and what better objects to consider than the collapsing objects of Oppenheimer and Snyder, whose endpoints were the pathological Schwarzschild solution? The second contributor to the revival was the discovery in 1963 of a new exact solution of Einstein's equations by the New Zealand–born relativist Roy P. Kerr. Kerr had been using a variety of sophisticated mathematical techniques that exploited symmetry principles to look for new solutions of the field equations. The solution he obtained was expressed in a rather obscure system of mathematical variables, and so when he gave a talk on this new solution at the First Texas Symposium on Relativistic Astrophysics in 1963 that I described in chapter 1, he must have seemed like a visitor from another planet to the astronomers and physicists who had not yet learned to communicate in the newly forming discipline. Nevertheless, in the discussion period following Kerr's talk, the Greek-born relativist Achilles Papapetrou admonished the audience to pay attention to this solution, because he had a feeling that it would one day prove to be important.

Papapetrou was right, for the Kerr solution turned out to be the unique solution for a rotating black hole, with the Schwarzschild solution being just a special case of the Kerr solution in which there is no rotation. Egged on by the problem of the quasars, relativistic astrophysicists spent the next ten years proving this and many other important features of the Schwarzschild and Kerr solutions. For instance, the pathological behavior of the Schwarzschild solution at the gravitational radius (and a similar behavior of the Kerr solution) proved to be purely a product of an inappropriate choice of mathematical variables. This did not alter the fact that the gravitational radius was special, however. The name "event horizon" was attached to the surface corresponding to this radius because it is a boundary for communication, just as the Earth's horizon is a boundary for our vision. An observer inside the event horizon cannot communicate with an observer outside by any means whatsoever, even by sending light signals. Nothing, not even light, can escape from the interior. On the other hand, light, matter, physicists, are all free to cross the event horizon going inward. This one-way property of the event horizon, allowing nothing to emerge, led John A. Wheeler, one of the fathers of the relativity revival, to coin the term black hole, during a 1967 conference in New York.

To an observer outside the horizon, the only feature of the black hole itself that is detectable is its gravitational field. (Any matter or radiation that remains outside the horizon, of course, is detectable.) Far away from the black hole, this gravitational field is indistinguishable from the gravitational field of any object of the same mass and angular momentum, such as a star. However, to an observer close to the horizon, things can be very unusual. The deflection of light can be so large that light can move on circular orbits, just outside the horizon (at 1.5 gravitational radii, for the Schwarzschild hole). For the

Kerr solution, rotation of the black hole produces the same dragging-of-inertial-frames effect whose detection is a goal of the Stanford gyroscope experiment (in the more mundane terrestrial setting), but if the observer goes close enough to the horizon, near the equator, the dragging of space-time becomes so strong that it is impossible for him to avoid being dragged around bodily with the rotation of the hole, no matter how hard he blasts his rockets to try to avoid it. These and many other physical and mathematical properties of black holes were established during a period of intense research by a score of theorists during the period from 1963 to 1974.

But instead of dwelling on the many unusual and remarkable properties of black holes, I will turn to the observational search for black holes, because that is where general relativity can play a practical role.

The first place you might think to look for black holes is in quasars, yet paradoxically, despite more than twenty years of intensive observation and theoretical analysis, the nature of quasars remains somewhat of a mystery. There is widespread agreement that the large red shifts in the spectra of quasars indicate that they are moving away from us at large velocities, and that, according to the picture of the expanding universe, they are therefore at very great distances. There is evidence that quasars were much more prevalent in the early universe than they are at present; as we look farther out in distance, we are also looking farther back in time because the light from the quasar takes a finite time to reach us, and it has been found that the number of quasars peaks at a time corresponding to an age of the universe about one-third of its present age. The powerhouse of the quasar is believed to be the active and violent central nucleus of a galaxy. The idea that this nucleus contains a relativistic collapsed object has changed little since 1963, when it was one of the main topics of the first Texas symposium. One of the most popular models for a typical quasar involves a supermassive black hole, weighing perhaps 100 million solar masses, at the center of the galactic nucleus (as large as this is, it may still be only 1/1,000 of the total mass of the galaxy). The black hole is gobbling up stars and gas at a ferocious rate, perhaps as much as 1 solar mass of material per year. As the material approaches the hole, friction from collisions with other material heats it up to temperatures high enough to make it radiate the enormous power we see on Earth. Although this picture is a crude fit to the observed characteristics of quasars, there are many details that must be ironed out before it can be accepted with real confidence. Some of these include accounting for the relative amount of energy emitted in various wavelengths, such as radio, optical, X rays, and so on; understanding the narrow jets of matter that can be seen shooting out on opposite sides of many quasars; and figuring out how such enormous black holes were formed in the first place.

This illustrates again why we can't expect to obtain quantitative tests of general relativity in such astrophysical contexts. The physical processes involving hot gas, whirling stars, and intense radiation in the nucleus of a galaxy are so messy that it will be impossible to learn anything precise about general relativity in quasars. Instead, we take the general relativistic description of the black hole as an assumption, and try to use it to build a model to make sense of the quasar observations. Before we become too sanguine about black holes, however, it is useful to keep in mind a comment that has been made by more than one cynic: "Whenever theoretical astrophysicists don't understand something, they invoke black holes."

However, there is at least one case closer to home, where we are on safer ground invoking a black hole. This case is the X-ray source Cygnus X1. The first astronomical X rays from sources other than the Sun were discovered in 1962, including the source Cygnus X1, the name denoting the first X-ray source in the constellation Cygnus. By 1967, about thirty such sources were known, all detected using instruments on sounding rockets and balloons launched far above the Earth's absorbing atmosphere. However, X-ray astronomy made a giant leap into the mainstream of astronomy with the launch of the Uhuru orbiting X-ray satellite in December 1970. The name Uhuru, meaning "freedom" in Swahili, was given to the satellite because it was launched from a facility in Kenya on that country's independence day. During its three-year lifetime, Uhuru charted more than two hundred X-ray sources. Later orbiting X-ray satellites found many many more sources, including ordinary stars, white dwarfs, neutron stars, galaxies, quasars, and a diffuse background of X rays, reaching us from all directions.

Uhuru's examination of the X rays from Cygnus X1 gave two crucial pieces of information that led to the conclusion that a black hole was present. The first was the observation that the X rays were variable in time in an irregular fashion, but on timescales less than 1 second. This meant that the region from which the X rays originated had to be no larger than the distance light could travel in 1 second, or 100,000 kilometers, in order for one side of the emitting region to know what the other side was doing. This implied that the object at the center of the X-ray emitting region had to be a compact object, such as a white dwarf, a neutron star, or a black hole, because a normal star, like our Sun, would have a diameter 10 times too large. The second piece of information provided by Uhuru was an accurate enough position for the source in the sky to make it possible to locate a star, known as HDE 226868, at the same location. Examination of the spectrum of light from this star showed that it was of a type that normally has a mass of between 12 and 20 solar masses, and further that it was in orbit about a companion. This was determined by looking at the Doppler shifts in the spectral lines, just as the orbit of the binary pulsar was determined by looking at the Doppler shifts of its pulse period. The companion had to be the X-ray source.

The model that emerged from these observations is one of a compact object and a massive star in orbit around each other. The tidal gravitational forces exerted by the compact object on the atmosphere of the star are strong enough to strip some of the atmosphere off the star, and pull it in toward the object. But because of the orbital motion of the pair, this gas does not fall straight toward the compact object; it misses it and goes into a circular, disk-shaped orbit around it. Because the gas in the disk closer to the object moves faster than the gas farther out, like the swirling drain in the swimming pool in chapter 11, there is friction between adjacent regions of gas. As a result, the gas is heated to temperatures sufficiently high to generate X-ray radiation. Also because of this friction, the gas slowly drifts inward, toward the compact object. So far so good, but still no black hole. The crucial point is that in order for the orbit of the normal star to have the velocities it was observed to have, the compact object must itself be moderately massive, at least 6 solar masses.

It is here that we apply general relativity to identify the compact object. It cannot be a white dwarf, because, as I remarked in chapter 10, the maximum possible mass for a white dwarf is about 1.4 solar masses, the so-called Chandrasekhar mass. This conclusion does not depend on general relativity, because white dwarfs are not very relativistic. What about a neutron star? General relativity plays an important role in the structure of neutron stars; nevertheless, relativists found a maximum possible mass for them as well, in this case about 3 solar masses, and by no means as large as 6. Therefore, it is not a neutron star. The only object left that can be massive enough to give the orbit of the companion, yet is small enough in size to allow the short-term X-ray fluctuations is a black hole.

Even though this argument is somewhat indirect, it has stood up to further observations of the system, as well as to attempts to propose alternate models that do not invoke black holes. Recently, an even more convincing black-hole candidate may have been found in a similar X-ray binary system in the nearby satellite galaxy called the Large Magellanic Cloud (appropriately enough, the source is called LMC X3). Several other binary X-ray systems are known, which are not believed to contain black holes; instead, they contain neutron stars as the compact object. In these systems, the inferred mass of the compact object is small enough to be compatible with the mass limits on neutron stars, and in some systems, the X-ray emission is supplemented by pulsed emission, indicating that the compact object is a pulsar, and therefore a neutron star. Thus, black holes are one of the many possible members of the astronomical family that may be found in different places. But because they have so few handles to grab hold of, we must use general relativity to infer their presence from a variety of clues.

The search for black holes continues in a variety of different ways. For example, a flight of the space shuttle in the spring of 1985 contained a telescope designed to look at the center of our galaxy in the infrared part of the spectrum, in order to search for evidence of what many astronomers think is a semi-supermassive black hole (only a million solar masses), surrounded by in-falling gas.

Although the search for black holes is yielding positive results, another search is not. This is the search for gravitational waves. True enough, the binary pulsar has confirmed that gravitational waves exist, and that the general relativistic description of orbital decay through gravitational-wave energy loss is quantitatively in accord with the observations. But the goal of detecting the waves themselves has not yet been attained.

Once gravitational waves have been detected, can we use them to test general relativity? Again, the answer is no, because the likely sources of gravitational waves are so complex that it will be impossible to disentangle the role of general relativity from the role of the messy source structure and dynamics. Instead, we will again take general relativity for granted, and use it to learn about the sources.

When this happens, we will be doing what relativists like to call gravitational-wave astronomy. Just as light-wave astronomers use optical, radio, X-ray, and infrared light, not to test Maxwell's theory of electromagnetism, but to study the universe, so will gravitational-wave astronomers use gravitational waves.

To see how this will come about, and to understand some of the difficulties involved, let us return to the simple description of a gravitational wave given in chapter 10, and refine it. We imagined a sphere being suddenly deformed into a cigar shape, producing a different gravitational-force field whose influence spread out from the body at the speed of light. But what is the nature of this varying force field? It could simply be a change in the gravitational force of attraction toward the body by some given amount over a finite region in space. But such a change would be unobservable, for the following reason. Let us put ourselves in a freely falling laboratory outside the body and look for a change in the force of attraction. Any change in the force will cause an acceleration that is the same for the laboratory as it is for us, according to the equivalence principle, and therefore, inside our laboratory, we will see nothing change. We and the laboratory will remain in free fall, with no apparent gravity inside, regardless of how the field changes. However, the tidal-force field exerted by the central body will change during the deformation. Remember that the tidal field is the difference in force or acceleration between one point and a nearby point in a region of finite size. It is what makes two particles separated by a finite amount horizontally in a freely falling laboratory fall toward each other, because they are actually falling toward the center of the external body. Now, the tidal gravitational field of a sphere is different from that of a cigar, so that even though we can't detect any change in the net acceleration of the laboratory, we can detect a change in the tidal force on any pair of particles that are separated by a finite amount in our laboratory.

This, then, is the essential nature of a gravitational wave: a wave that travels with the speed of light that causes separated bits of matter to feel a force relative to each other. It turns out that the force is always transverse or perpendicular to the direction of propagation of the wave.

This discussion also allows us to summarize the kinds of sources that might produce gravitational waves: they are sources in which mass changes its shape or moves around. Therefore, a vibrating star whose shape varies is a source. So is a binary-star system (as we saw in chapter 10), so is a black hole swallowing a star, two stars or black holes colliding or whizzing past each other, a collapsing rotating star (part of a supernova), and so on.

There is another important feature of gravitational waves, and this is the discouraging part: their strength. While electromagnetic radiation is strong enough to have many practical (and impractical) applications, gravitational radiation is extraordinarily weak. One of the strongest imagined sources of gravitational waves, a rotating star that collapses to form a black hole in our galaxy, will produce a field of tidal forces that will cause two masses separated by 1 meter to move together and apart by only one-hundredth of the diameter of an atomic nucleus! For sources farther away, the effect will be even weaker, because the size of the tidal-force field falls off as the inverse distance to the source.

The goal of gravitational-wave searchers is to detect these miniscule forces. The first searcher, and the pioneer of gravitational-wave astronomy, was Joseph Weber. Born in Paterson, New Jersey in 1919, and educated at the U.S. Naval Academy and the Catholic University of America, he joined the faculty of the University of Maryland in 1948, and has been there ever since. An accomplished experimentalist with a deep appreciation for theoretical principles, he was one of three research groups in the early 1950s that independently enunciated the basic ideas that led to the development of the maser. Around 1958, Weber began working on the problem of how to detect the tiny tidal forces associated with a gravitational wave, first doing the required theoretical calculations to determine just what the physical effects of a passing wave would be, and then building an apparatus. By 1965, he had put a simple detector into operation.

Weber's basic concept for a gravitational-wave detector is still in use today. It consists of a solid cylinder, usually of aluminum (the reason for aluminum is a mundane one: it is cheap), with a typical size of one-half a meter in diameter by 1 or 2 meters in

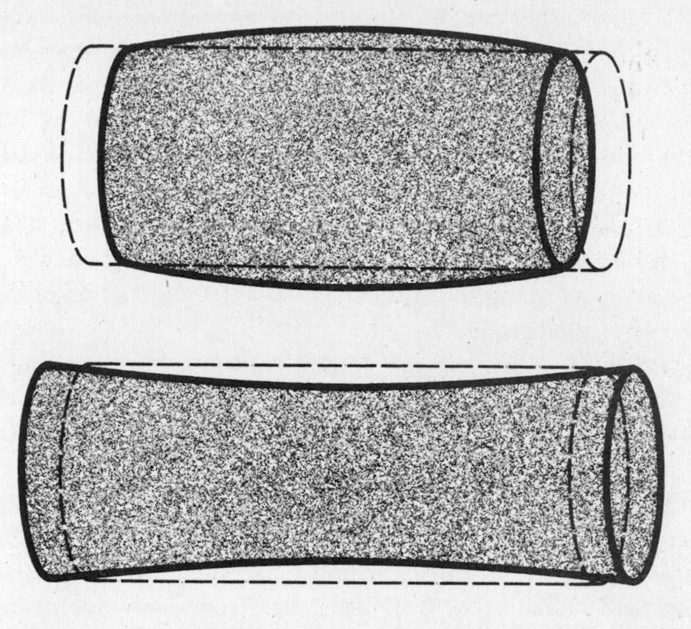

Figure 12.1 Effect of gravitational wave on a cylinder of aluminum. Wave is traveling out of the page. First, tidal forces of the wave compress the ends while expanding the midsection, then expand the ends while compressing the midsection. There are no forces acting parallel to the direction of the wave, that is, perpendicular to the page.

length. Its weight is about 1.5 tons. When a gravitational wave passes the cylinder in a direction perpendicular to its axis, the tidal force in the wave tries to move the two ends of the bar toward each other and away from each other (see figure 12.1). Because of the material in the bar between the ends, they do not move freely; instead, the forces in the wave make the ends oscillate at a frequency characteristic of the size and shape of the bar. A useful analogy is that of two balls connected together by a spring. When the balls are pushed together or pulled apart and then released, the restoring force of the spring will make them continue to oscillate afterward, until internal friction damps out the vibrations. Therefore, the passing gravitational wave sets the bar into an end-to-end oscillation, with a frequency that, for bars of the previously mentioned dimensions, is typically thousands of cycles per second, in other words, in the kilohertz band. The compression and extension of the bar will produce strains near the midpoint of the bar, and Weber attached devices around the bar at the middle to convert the strains into electrical signals that could be recorded and analyzed. Later detectors designed by other researchers had devices attached to the ends that could measure the motion of the two end faces directly.

In 1968, Weber made the stunning announcement that he was detecting signals simultaneously in two detectors spaced 1,000 kilometers apart, one in Maryland, and the other at Argonne National Accelerator Laboratory near Chicago. The reason for using two detectors is simply that any one detector is often in oscillation because of disturbances from the environment (seismic noise, trucks rolling by, and so on) despite sophisticated attempts to isolate the bar from such noise, and because of the inevitable random internal motions by the atoms inside the bar produced by heat energy. As I pointed out earlier, the motion of the bar produced by a gravitational wave is at most a fraction of the size of an atomic nucleus, so the expected signal is exceedingly small. Therefore, in a single bar, it is difficult if not impossible to distinguish a disturbance from a gravitational wave from a disturbance of an environmental or thermal origin. As early as 1967, Weber had reported disturbances in a single bar, but he could not reliably claim that they were from gravitational waves. However, with two detectors separated by such a large distance, a disturbance that appears simultaneously in both detectors would be unlikely to be environmental or thermal because the probability of such a random coincidence is very small. Coincident events would therefore be good candidates for gravitational waves. Even more remarkable than the 1968 report of coincident events was his announcement in 1970 that the rate of such events was highest when the detectors were oriented perpendicular to the direction of the center of the galaxy, implying that the sources were indeed extraterrestrial, perhaps concentrated near the galactic center. These reports caused a sensation both in scientific circles and in the popular press.

There were two problems, however. The observed events occurred with a disturbance size and at a rate (around 3 times per day) that shocked theorists, for it implied a number of bursts of gravitational waves at least 1,000 times what they predicted. This in itself was not necessarily bad, for often the mark of an important experimental discovery in physics is the degree to which it upsets theoretical sacred cows. The second problem was more telling, however. The main body of Weber's coincidence results were reported between 1968 and 1975. But by 1970, independent groups had built their own detectors with claimed sensitivities equal to or better than Weber's, yet between 1970 and 1975, none of these groups saw any unusual disturbances over and above the inevitable noise.

Weber's reported detections are now generally regarded as a false alarm, although there is still no good explanation for the coincidences if they were not gravitational waves. Nevertheless, they did leave behind an important legacy. Weber's experiments initiated the program of gravitational-wave detection, and inspired other groups to build better detectors. One of the main improvements has been to cool the entire bar and associated sensing devices to one or two degrees above absolute zero, in order to reduce the size of the disturbances due to the thermal motions of the atoms inside the bar. A dozen laboratories around the world are engaged in building and improving upon the basic "Weber bar" detector, some using bigger bars, some using smaller bars, some working at room temperature, and some working near absolute zero, some using aluminum, some using other materials such as sapphire that might have an improved response to the gravitational-wave excitations. Several other groups are working on a radically different scheme. Known as a laser interferometer, this device takes light from a laser, splits it into two beams traveling at right angles to one another, reflects the beams from two mirrors that return the beams to the starting point, and combines them in such a way that they interfere with each other. A passing gravitational wave will change the length of one arm of the apparatus relative to the other, and cause the interference pattern to vary. The builders of both types of "second generation" detectors claim sensitivities between 1,000 and 10,000 times better than Weber's detectors of 1968–70.

Yet they still have a long way to go. The current detectors are capable of detecting gravitational waves from a decent supernova in our galaxy. There is only one problem. The last known supernova in our galaxy was in 1604. No one wants to wait around for three hundred years to get a signal in his detector. Thus, the ultimate goal is to improve the detectors by another factor of around 10,000, in order that they can detect gravitational waves from other galaxies. These waves will be weaker as seen on Earth because they have to come from farther away, but if as many as 1,000 galaxies are included, the rate of interesting source events such as supernovas may become more reasonable, say one per month.

So the program started by Weber a quarter of a century ago is far from over, and when it achieves its first successful detection, it will initiate an exciting new kind of astronomy. It is in this new gravitational-wave astronomy that general relativity will be an important and useful tool. For a given source, whether it is a collapsing star or a binary-star system, general relativity makes a definite prediction of the size, frequency spectrum, and duration of the gravitational waves emitted. One of the goals of applied general relativists is to build a catalogue of possible sources with the associated gravitational-wave characteristics, so that when the first waves are ultimately detected, it may be possible to learn something about the source from the characteristics of the wave (see figure 12.2). For example, a binary-star system emits steady gravitational waves, primarily at a frequency equal to twice the orbital frequency, with an amplitude that is related to the mass and separation of the bodies. On the

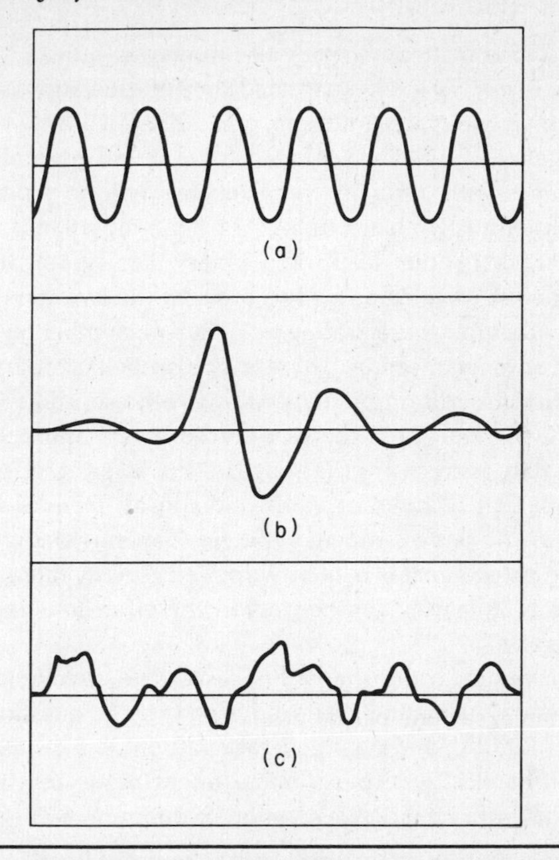

Figure 12.2 Gravitational waves from different sources. (a) Binary-star system: regular waves with a well-defined period. (b) Supernova collapse to a black hole: large burst of waves followed by decaying smaller waves from jiggling of the black hole. (c) Gravitational-wave background: random noisy background from many sources or from the big bang.

other hand, the collapsing rotating core of a supernova emits a burst of waves during the final fraction of a second of the collapse, when the matter is moving at a significant fraction of the speed of light, and is compressed down to neutron-star or black-hole sizes. This is because, roughly speaking, the more the mass involved, the higher the speed of the mass, and the more compressed the system, the stronger the gravitational waves. If the matter collapses all the way to a black hole, the black hole may be left in an excited state of oscillation, a jiggling ball of invisible space-time jello, and will emit a train of waves with a set of well-defined frequencies that will gradually damp out. Waves with different characteristics will be produced by the collision of two black holes, or by the vibrations of a neutron star, or by the death of a binary-star system in which gravitational-wave damping has brought the two stars so close together that they are spiraling in toward each other on the way to a cataclysmic meeting. There may also be a significant universal background of gravitational-wave noise, some left over from the big bang, and some produced by the sum total of all the myriad sources throughout the universe. Using all the mathematical techniques at their disposal, from pencil and paper, to the largest supercomputers, general-relativity theorists are using general relativity applied to gravitational-wave astronomy as simply another tool in the effort to comprehend the universe.

We have now come almost full circle. The discoveries of the astronomers during the 1960s made us face the question, "Was Einstein Right?," and now general relativity is a central tool in the astronomical enterprise. Some might argue that we have taken this most remarkable creation of the human mind, which has revolutionized our concept of space, time, and the universe, and reduced it to the mundane level of just another part of the apparatus of science.

But I wouldn't agree with that assessment. If general relativity were devoid of applications in physics or astronomy, it would be a closed subject, and would soon return to the stagnation and sterility that characterized it before the relativistic renaissance of recent decades. It is precisely those applications in black-hole searches, gravitational-wave astronomy, and cosmology that have imbued the subject with a vitality and a vigor that have made it an open-ended field, not a closed book, despite the fact that the structure of the theory has not changed in seventy years.

Yet we still had to find out whether indeed the theory was right. No matter how revolutionary it was, no matter how beautiful its structure, our guide had to be experimentation. Equipped with new measuring tools provided by the technological revolution of the last twenty-five years, we put Einstein's theory to the test. What we found was that it bent and delayed light just right, it advanced Mercury's perihelion at just the observed rate (solar oblateness notwithstanding), it made the Earth and the Moon fall the same, and it caused a binary system to lose energy to gravitational waves at precisely the right rate. What I find truly amazing is that this theory of general relativity, invented almost out of pure thought, guided only by the principle of equivalence and by Einstein's imagination, not by a need to account for experimental data, turned out in the end to be so right.

Appendix

Special Relativity: Beyond a Shadow of a Doubt

IT IS DIFFICULT to imagine life without special relativity. Just think of all the phenomena or features of our world in which special relativity plays a role.

Atomic energy, both the explosive and the controlled kind. The famous equation $E = mc^2$ tells how mass can be converted into extraordinary amounts of energy.

Chemistry, the basis of life itself. The chemical laws are determined by the fact that electrons in the atom are arranged in shells (this is the origin of the periodic table of the elements). The fundamental principle that limits the number of electrons that can occupy a given shell is a product of the marriage of special relativity with quantum theory.

Evolution of the species. One possible source of the genetic mutations that permit evolution of living species is cosmic rays. At sea level, the main component of the cosmic rays is the unstable particle known as the mu meson or muon. But the muon is so unstable that it would decay long before reaching sea level from the upper atmosphere where it is created in the collision of an extraterrestrial cosmic-ray proton with an atom, if it weren't for the time dilation of special relativity, which increases its lifetime as a consequence of its high velocity.

The U.S. National Budget. In 1983, particle physicists proposed that the United States build a gigantic new particle accelerator called the superconducting super collider. The machine would be a ring large enough that, if it were built in Washington, D.C., it could completely encircle the city and much of its suburbs. It would cost around 3 billion dollars. One reason for the enormous size and cost is the special relativistic increase in the inertia of a particle moving near the speed of light that makes it harder and harder to accelerate it to higher velocities.

Special relativity is so much a part not only of physics but of everyday life, that it is no longer appropriate to view it as the special "theory" of relativity. It is a fact, as basic to the world as the existence of atoms or the quantum theory of matter. It has been tested time and time again by experiments designed to check its predictions directly. Perhaps more importantly, however, it has also been tested and confirmed, albeit indirectly, by the fact that the welding together of special relativity with the quantum theory has led to a total understanding of atomic physics, electromagnetism, and nuclear physics, and has provided the foundation for modern advances in elementary particle physics.

That such a powerful and far-reaching conception should be based on a few simple premises is truly remarkable. It is a product of Einstein's genius—taking a commonplace observation, combining it with some simple imaginary experiments, and arriving at a revolutionary conclusion. He did this with special relativity, and he also did it with general relativity.

Special relativity is based on two concepts, the inertial frame, and the principle of relativity. An inertial frame of reference is any region, such as a freely falling laboratory, or a laboratory far from the Earth in interstellar space, in which all objects move in straight lines with constant velocity. The idea here is to get rid of gravity as much as possible, at least for the time being. How Einstein incorporated gravity into relativity is the subject of chapter 2.

However, for the purpose of illustration, it is sometimes useful to consider as a partially inertial frame, a laboratory on the surface of the Earth, so long as we do not deal with motion in the vertical direction, where it is clear that gravity makes bodies move on curved paths or with changing velocity. But if we consider only horizontal motion, such as that of a rolling ball, we can make good use of such frames. One example might be a frame at rest on the Earth, in which two observers, separated by 5 meters, roll a ball toward each other with a velocity of 1 meter per second, and measure the time taken to cross the distance. Another example might be a frame on a train moving at 10 meters per second, in which two observers, also separated by 5 meters, roll an identical ball at 1 meter per second toward each other.

The second ingredient of special relativity is the "principle of relativity," which states that the result of any physical experiment performed inside a laboratory in an inertial frame is independent of the velocity of the frame. In other words, the laws of physics must have the same form in every inertial frame. Another, less precise way of saying this is that all inertial frames are equivalent. In the rolling ball example, the principle of relativity would demand that in each laboratory, the time taken by the rolling ball should be 5 seconds. Identical physical setup and identical laws imply identical outcomes.

So far, nothing I have said is unusual or surprising and is exactly the same as the standard picture of Newtonian mechanics that held sway for over two centuries. Einstein's revolutionary insight was to demand that the principle of relativity apply to all the laws of physics, not just to the laws of mechanics. He demanded that the principle should also apply to the laws of electromagnetism. Why was this so revolutionary?

It was revolutionary because of the speed of light. In the rolling ball example, the speed of the rolling ball was a quantity that was determined by the observers; it is what physicists call an initial condition. Now, we all know what happens if an observer on the ground watches the rolling-ball experiment on the moving train. He sees a ball being rolled at 1 meter per second on a train that is moving at 10 meters per second; therefore the speed of the ball according to him is 11 meters per second. The initial condition of the ball on the train is different for the ground and for the train observers. As it turns out, of course, the ground observer finds that it takes the same amount of time for the ball to roll from one train observer to the other. Despite the ball's larger speed as seen from the ground, the train observer toward whom it is rolling is also moving away at 10 meters per second. The additional ground distance it covers in catching up compensates for its higher speed, so the time taken is the same 5 seconds. This result is in accord with the Newtonian view of an absolute time that is the same for all observers. We will see that things are very different in special relativity.

What makes light different? According to the theory of electromagnetism developed in the 1870s by James Clerk Maxwell (1831–79), the speed of light was a fixed constant of nature, given approximately by 299,800 kilometers per second. In Einstein's view, this speed was an integral part of the laws of electromagnetism and was not an initial condition like the speed of the rolling balls. Therefore, according to Einstein's principle of relativity, the speed of light should be the same in every inertial frame. If two observers on the ground send a light signal to one another, they should get the standard value, usually denoted by c. Two observers on a moving train should get the same value when they send a light signal to one another. But in addition, two observers on the train should get the same value when they measure the speed of a light signal sent between the two observers on the ground, regardless of the fact that the ground observers are moving relative to the train. The ground observers should similarly get the same speed for a light signal sent on the train.

This idea flew in the face of turn-of-the-century physical intuition, which viewed the speed of light only partially as a fundamental law and partially as an initial condition. It was true that Maxwell's equations prescribed a fixed value for this speed, but physicists postulated that this was the speed relative to a medium known as the aether. The aether was supposed to fill space and to be at rest with respect to the entire universe. Therefore, just as in the case of an observer on a train watching rolling balls on the Earth, an observer moving relative to the aether would find that the speed of light would differ from c. For instance, if the observer moved parallel to the light ray in the opposite direction, he would find a speed of c plus his own velocity; if he moved perpendicularly to the light ray, he would find a speed given by the Pythagorean theorem, namely the square root of the sum of the squares of c and of his speed, and so on.

The seeds of the downfall of this idea were planted in 1887 by two U.S. scientists, Albert A. Michelson (1852–1931) and Edward W. Morley (1838–1923), who performed an experiment designed to detect the motion of the Earth through the aether. The experiment used an apparatus called an interferometer, originally designed by Michelson to make accurate measurements of the speed of light (see figure A.1). Schematically, Michelson's interferometer consists of two straight arms set at right angles to each other. Each arm has a mirror at one end. At the intersection where the arms are joined, a half-silvered mirror splits a light beam into two, each traveling down one arm, each reflected back by the mirror at the end of each arm. When the two beams are recombined, they interfere in such a way as to produce a characteristic pattern of fringes that depends on the difference in time required for the two beams to make the round trip. If one arm of the interferometer is parallel to the Earth's supposed motion through the aether, the speed in one direction is less than c by the amount of the Earth's speed

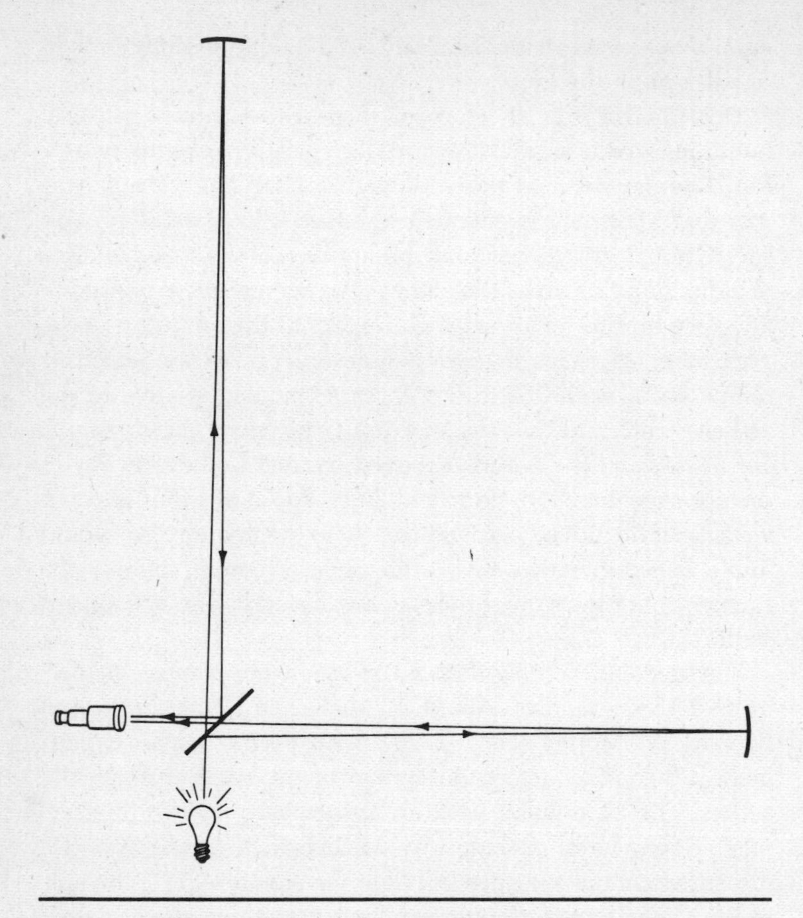

Figure A.1 Michelson interferometer. Light from source is split by a half-reflecting mirror, and travels along two perpendicular arms. Mirrors at each end reflect beams back. Beams are brought together, and pattern of light and dark interference fringes is observed.

and on the return leg is greater than *c* by the same amount. Along the other leg, perpendicular to the motion, the speed is the same out and back, but with a value given by the square root of the sum of the squares of the two speeds. If the apparatus is rotated through 90°, the roles of parallel and perpendicular arms are reversed, so the pattern of interference fringes should shift (the trick of rotating the apparatus eliminates the need to make the arms exactly equal in length, a difficult thing to achieve in practice). To avoid distortion of the arms of the instrument by strains and vibrations, Michelson and Morley mounted it on a concrete block floating in a bath of mercury, and observations were made as it was rotated slowly and continuously about a vertical axis. If the speed of light were to vary because of the Earth's velocity around the Sun, which is about 30 kilometer per second, the expected shift in the interference pattern would be four-tenths of a wavelength of the light they were using. However, no shift as large as four-hundredths of a wavelength was observed. As far as the moving Earth was concerned, the speed of light was the same along both arms, no matter what.

This is just what Einstein's principle of relativity demands: The speed of light should be the same in all inertial reference frames, whether at rest with respect to the universe, or at rest with respect to the Earth and no matter what the state of motion of the source of the light. The idea of an aether is totally superfluous. It is interesting to note, however, that Einstein himself was apparently unaware of the specifics of the Michelson-Morley experiment when he put forward the theory of special relativity in 1905, although he did allude vaguely to experiments that had failed to detect motion through the aether.

The notion that the speed of light is the same is really the key to understanding the important physical consequences of special relativity. The first important consequence is that what we mean by "simultaneous" may depend on our frame of reference. Consider two observers on the ground, equidistant from a central master observer, and on opposite sides (see figure A.2). The two wish to synchronize their clocks, so they agree beforehand that when each receives a light flash emitted in all directions from the master observer, he will set his clock to a prescribed time. Because they are equidistant from the master observer and the speed of light is the same in both directions, they naturally assume that they receive the signals simultane-

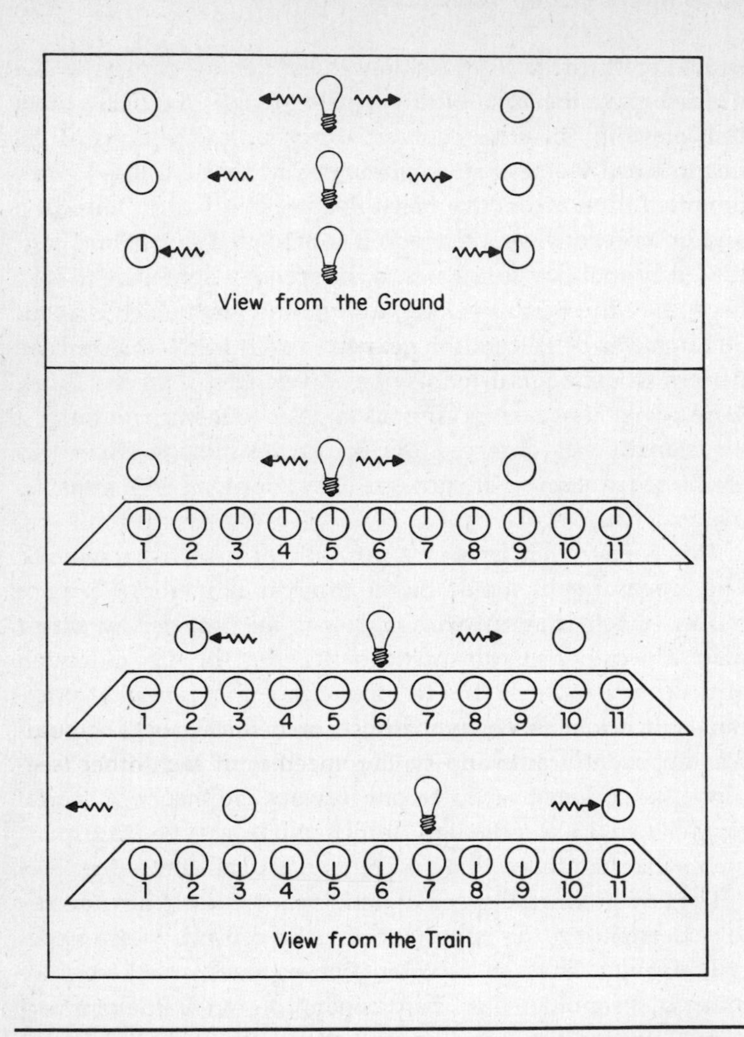

Figure A.2 Simultaneity. Top: View from the ground. Master light source midway between two observers emits a light flash. Each observer sets his clock to read the same agreed-upon amount when flash is received. They agree the reception of the flashes was "simultaneous." Bottom: View from a train moving past at one-third the speed of light. Clocks on the train have previously been synchronized using a similar technique on the train. Speed of the light flash as seen by the train observers is the same as seen by the ground observers. Leftward moving signal therefore reaches ground observer "before" rightward moving signal, because it has less distance to travel. Thus, two events seen to be simultaneous on the ground are seen to be not simultaneous on the train.

ously. Look at what happens, however, from the point of view of a set of observers on a moving train who watch the same flash of light. The master observer emits his signal, which is seen to move with the same speed in both directions despite the motion of the train (principle of relativity). However, while the signals are propagating, both ground observers are moving relative to the train, so the signal sent in the forward direction has less distance to cover before it intersects the forward observer, while the signal sent in the backward direction has a greater distance to cover in catching up to the receding observer. Thus, according to the train observers, the signal is received by the forward observer before it is received by the rear observer; whereas, in the ground frame, the two observers viewed the receipts as simultaneous. This conclusion helped abolish the previously accepted Newtonian concept of an absolute, universal time that was the same for all observers.

The destruction of absolute time was completed by the other important relativistic effect on time known as time dilation. According to this effect, a clock moving through an inertial frame will tick more slowly than a set of identical clocks laid out throughout the frame and synchronized with each other (see figure A.3). The reason for having many clocks in the inertial-frame laboratory is that one can only compare clocks with each other unambiguously when they are side by side (otherwise you have to send light signals between them and worry about what happens during propagation of the signals), so the moving clock will have to be compared with different clocks at different places along its path. But as long as the laboratory clocks are identical and synchronized, this should be no problem. Suppose, for simplicity's sake, that all our clocks are made in the following way: Two mirrors are placed at the ends of a rod, and the rod is placed perpendicular to the direction of motion of the moving clock. A light signal is reflected back and forth between the two mirrors, and each time the signal returns to the first mirror, a tick is recorded. Given the length of the rods and the speed of light, we can determine the time interval

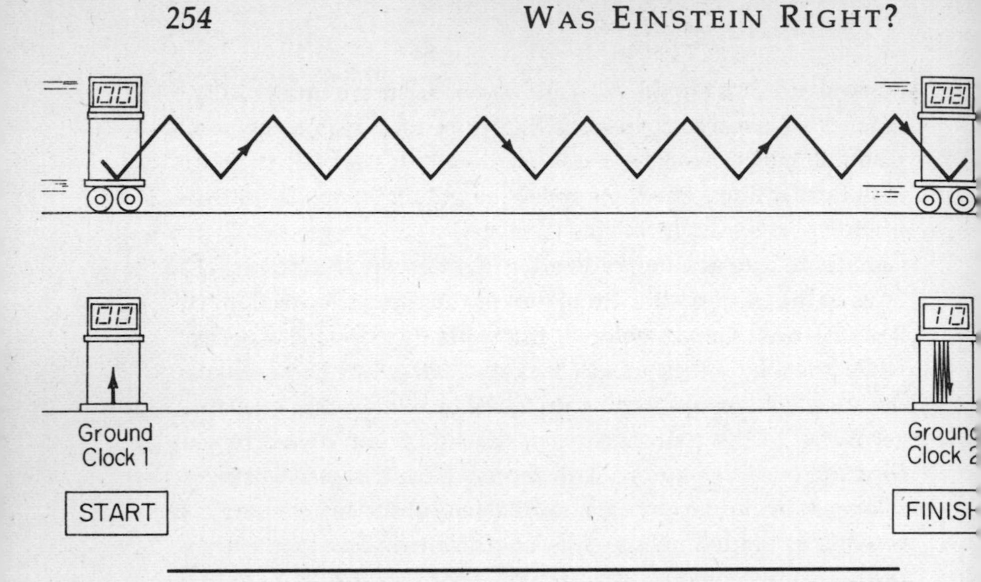

Figure A.3. Time dilation. Observers use clocks consisting of parallel mirrors mounted on floor and ceiling of a laboratory. Each time light signal hits the floor, a tick is recorded. Moving clock travels at three-fifths of the speed of light past two synchronized ground clocks constructed in the same way. It takes ten ticks to do so, as measured by the ground clocks. As seen from the ground, the reflected light signal of the moving clock traces a saw-tooth pattern, so between reflections, it must cover a larger distance than the height of the laboratory. Because the speed of the light signal is the same as in the ground laboratory, it takes longer for the light to cover the added distance, so only eight ticks are recorded by the moving clock. Thus, the moving clock ticks more slowly than the stationary clocks.

between ticks for all our laboratory clocks, and we would get the same value, of course. An observer moving along with the moving clock would also get the same value, because to him the speed of light is the same, and the length of the rod is the same. But what is the time between ticks of the moving clock as seen by the laboratory clocks? As seen by the laboratory observer, the reflected light signal of the moving clock traces out a saw-toothed pattern, so that the distance it has to travel between reflections is a little larger than if the clock had been at rest, again by the Pythagorean theorem. The speed of the light signal is still c, of course. Therefore the time, according to the laboratory clocks, for each tick of the moving clock is a little longer than 1 tick, so the moving clock must be running slower. As the moving clock passes successive laboratory clocks, it is seen to fall farther and farther behind.

This simple argument can be used to calculate the correct mathematical formula for the time dilation. But it turns out that the effect is universal and does not depend on the specific nature of the clocks, so all moving clocks experience this effect, whether they are atomic clocks, unstable elementary particles, or people governed by biological clocks.

The observational evidence for time dilation is overwhelming. As I have already mentioned, it is this time dilation that slows down the decay rates of the unstable muons generated by cosmic rays, allowing them to reach sea level. Quantitative tests of time dilation have been performed numerous times at particle accelerators. A classic experiment was performed in 1966 at the accelerator at the European high-energy laboratory in Geneva, called CERN. Muons produced by collisions at one of the targets in the accelerator were deflected by magnets so that they would move on circular paths, and so could be stored for a decent interval (such "storage rings" have become an integral part of most high-energy accelerators). Their speeds were 99.7 percent of the velocity of light, and the observed twelve-fold increase in their lifetimes agreed with the prediction with 2 percent accuracy. Modern-day atomic clocks keep time so accurately that in order to compare time kept on Earth at such installations as the U.S. Naval Observatory or the Bureau International de l'Heure in Paris with time kept on atomic clocks in orbiting satellites, the time dilation must be accounted for, or discrepancies will arise between the various observatories. (Another effect, known as the gravitational red shift, a consequence of curved space-time, must also be taken into account. This effect is the subject of chapter 3.)

Another effect that is a consequence of special relativity is known as the Lorentz-FitzGerald contraction, an apparent shortening of the length of a moving rod as measured by an identical rod at rest. This effect was first proposed by the Irish physicist George F. FitzGerald in 1892, and expanded upon by the Dutch theorist Henrik A. Lorentz in 1895, in their attempts to understand the Michelson-Morley experiment and the nature of electromagnetism without discarding the idea of an aether. But the Lorentz-FitzGerald contraction is a natural outcome of the principle of relativity and can be derived using rods and the same mirror-clocks used before to understand the time dilation. On the other hand, it is a rather difficult effect to see experimentally, because it is hard to accelerate macroscopic rods to high enough velocities to make the effect noticeable. It is useful to note that for the effects I have been describing, the deviations from our Newtonian intuition can be estimated by taking the ratio of the velocity of the object or frame to the speed of light, and squaring it. For an automobile moving at 100 miles per hour, this quantity is twenty parts in a million billion, or two-trillionths of a percent.

The principle of relativity can also be applied to more complicated situations, such as a collision between two bodies, or the motion of a charged body in an electric field. In order for the outcomes of experiments like these to be independent of the inertial frame, the effective mass, or inertia, of a moving particle must increase. This relativistic increase of inertia is what prevents particles from being accelerated up to and beyond the speed of light, because the inertia of the particle increases without bound as its speed approaches *c*. This has been observed countless times in particle accelerators, and must be figured into all the engineering specifications and cost estimates for more powerful accelerators.

Almost as a corollary to this effect, Einstein derived one of the most famous (and sometimes notorious) equations in all physics. He first demonstrated that the increase of inertia of a moving particle is equivalent to an increase in its energy. However, in a frame moving along with the particle, the particle is at rest, so all it has is mass, usually called the "rest mass." But because the two inertial frames are equivalent, what we call mass in one frame and what we call energy in another frame must really be the same thing, so mass and energy are really equivalent. The conversion factor between mass and energy was just the square of the speed of light: hence the famous equation $E = mc^2$. The explicit validation of this equation, both in nuclear power plants and in atomic weapons, has provided at the same time a possible key to the survival of civilization and the tools for its destruction.

In some ways, however, the most important consequences and confirmations of special relativity arise when it is merged with the quantum theory. This merging, first carried out by the British theorist Paul A. M. Dirac in 1928, led to numerous predictions in agreement with experiments. These include the fact that electrons (indeed all elementary particles) are endowed with a property called "spin," and the prediction of the existence of antimatter. The merging also produced new equations of atomic structure leading both to an understanding of the electronic shells that govern chemistry and to a correct picture of the details of atomic spectra. The simultaneous merging of special relativity with quantum mechanics and electromagnetism led to one of the most important and successful physical theories of the twentieth century, called quantum electrodynamics (QED), which provides a complete accounting of the interactions between charged particles. The structure of QED now provides the model for the latest theories of the elementary particles, based on quarks and other exotic particles. Even the name of this new class of theories, quantum chromodynamics (QCD) is modeled after QED. The stunning achievements of these theories of the fundamental particles and forces would have been impossible were it not for the underlying structure provided by special relativity.

Suggestions for Further Reading

Popular and General Interest Books

Calder, N. *Einstein's Universe.* Penguin Books: New York, 1981. Survey of general relativity, black holes, and cosmology. Based on the television production.

Davies, P. C. W. *The Search for Gravity Waves.* Cambridge University Press: Cambridge, England, 1980. Discussion of the nature of gravitational waves, how they are generated, how they can be detected, and the status of the search as of 1980.

Douglas, A. V. *The Life of Arthur Stanley Eddington.* Thomas Nelson & Sons: London, 1956. Biography of Eddington, including excerpts from his notebook on the day of the 1919 eclipse.

Einstein, A. *Relativity: The Special and General Theory.* Crown: New York, 1961. English translation of the 1917 popular book by Einstein, who may have joked at the time that only twelve people could understand it.

Gardner, M. *The Relativity Explosion.* Vintage: New York, 1976. Revised and updated version of the 1962 classic, *Relativity for the Million.*

Greenstein, G. *Frozen Star: Of Pulsars, Black Holes and the Fate of Stars.* Freundlich Books: New York, 1984. Award-winning account of nature and discovery of compact or collapsed objects in the universe.

Johnson, P. *Modern Times: The World from the Twenties to the Eighties.* Harper & Row: New York, 1983. History of the world starting with the 1919 eclipse expeditions. The author assesses the influence of moral "relativism" on twentieth-century events.

Mermin, N. D. *Space and Time in Special Relativity.* McGraw-Hill: New York, 1968. Elementary but thorough exposition of special relativity, with some algebra.

Pagels, H. R. *Perfect Symmetry: Search for the Beginning of Time.* Simon & Schuster: New York, 1985. Unification in particle physics, and its connection with cosmology.

Pais, A. *'Subtle is the Lord . . .': The Science and the Life of Albert Einstein.* Oxford University Press: New York, 1980. Beautiful biography of Einstein the man and his work. Discussion of the physics quite technical in places.

Weinberg, S. *The First Three Minutes: A Modern View of the Origin of the Universe.* Basic Books: New York, 1977.

Popular Articles on Selected Topics

Chaffee, F. H., Jr. "The Discovery of a Gravitational Lens." *Scientific American* (November 1980):70.

Faller, J. E., and Wampler, E. J. "The Lunar Laser Reflector." *Scientific American* (March 1970):38.

Fisher, A. "Testing Einstein (again) with a Relativity Satellite." *Popular Science* (August 1983):55. The gyroscope experiment.

Gale, G. "The Anthropic Principle." *Scientific American* (December 1981):154.

Shapiro, I. I. "Radar Observations of the Planets." *Scientific American* (July 1968):28.

Weisberg, J. M., Taylor, J. H., and Fowler, L. A. "Gravitational Waves from an Orbiting Pulsar." *Scientific American* (October 1981):74.

Will, C. M. "Gravitation Theory." *Scientific American* (November 1974):24.

Textbooks on Special and General Relativity

Eddington, A. S. *The Mathematical Theory of Relativity.* Cambridge University Press: Cambridge, England, 1923. One of the first textbooks on relativity.

French, A. P. *Special Relativity.* Norton: New York, 1968. Introductory college-level text on special relativity.

Misner, C. W., Thorne, K. S., and Wheeler, J. A. *Gravitation.* Freeman: San Francisco, 1973. Includes extensive discussion of experimental tests of general relativity.

Schutz, B. F. *A First Course in General Relativity.* Cambridge University Press: Cambridge, England, 1985.

Weinberg, S. *Gravitation and Cosmology: Principles and Applications of the General Theory of Relativity.* Wiley: New York, 1972.

Technical Books on Special Topics

Bertotti, B., ed. *Experimental Gravitation.* Academic Press: New York, 1974. Lectures given in 1972 at a summer school devoted to this topic. A wealth of experimental detail.

Dicke, R. H. *The Theoretical Significance of Experimental Relativity.* Gordon & Breach: New York, 1964. The status of experimental relativity as Dicke saw it in 1964.

Ehlers, J., Perry, J. J., and Walker, M. *Ninth Texas Symposium on Relativistic Astrophysics.* New York Academy of Sciences: New York, Vol. 336, 1980. Contains first report of detection of gravitational radiation using the binary pulsar.

Lorentz, H. A., Einstein, A., Minkowski, H., and Weyl, H. *The Principle of Relativity.* Dover: New York, 1952. Translations of original articles on relativity. Einstein's papers on special and general relativity can't be beaten for clarity and elegance.

Robinson, I., Schild, A., and Schucking, E., eds. *Quasi Stellar Sources and Gravitational Collapse.* University of Chicago Press: Chicago, 1965. Proceedings of the First Texas Symposium on Relativistic Astrophysics in 1963. Fun to compare with proceedings of the ninth.

Roseveare, N. T. *Mercury's Perihelion from Le Verrier to Einstein.* Clarendon Press: Oxford, England, 1982. Detailed description of alternative hypotheses proposed to explain the anomalous advance in Mercury's perihelion.

Wesson, P. S. *Cosmology and Geophysics.* Adam Hilger: Bristol, 1978. Discussion of variable-G theories, the large numbers hypothesis, and observational evidence.

Will, C. M. *Theory and Experiment in Gravitational Physics.* Cambridge University Press: Cambridge, England, 1981. Mathematical tools needed to calculate effects of general relativity and alternative theories and to analyze experiments. Includes a comprehensive list of references as of 1979 to original papers for theoretical developments and experiments described in this book. For a summary and references updated to 1983, see Will, C. M. "The Confrontation Between General Relativity and Experiment: An Update." *Physics Reports* 113 (1984):345.

Index

Abbott, E. A., 24
aberration, 71
absolute time, 248, 253
absolute zero, 223, 239
action and reaction, law of, 130
action at a distance, 23
Adams, John Couch, 90
aether, 67, 249, 251, 256; *see also* special relativity
Alley, Carroll O., 145
ALSEP, 144
alternative theories of gravity, 13, 40, 137, 149; and space curvature, 74, 117; *see also* Brans-Dicke theory
aluminum: in gravitational-wave detectors, 236
anchored spacecraft, 130, 131, 132
Andromeda galaxy, 162
Angstrom, 51
angular momentum: in Earth-Moon system, 174
Annalen der Physik, 79
Annals of Physics, 5, 9
anthropic principle, 167–70
anti-relativity program, 78
anti-Semitism, 78
antigravity, 166
antimatter, 257
apastron 188, 192, 193, 199
aphelion, 104
apogee, 44
Apollo 11, 144, 145
Apollo 14 and 15, 145
Apollo Lunar Surface Experiments Package (ALSEP), 144
Apollo program, 14, 140, 144; *see also specific missions*
arcsecond, 67
Arecibo radio telescope, 119, 128, 181, 184, 186, 189, 206
Argonne National Accelerator Laboratory, 238
Aristotle, 27, 161
Armstrong, Neil, 144
Arrow to the Sun, *see* Starprobe
asteroids, 179
astrometry, 67, 102
astronomical unit, 4; error in, 4, 109
astronomy: and usefulness of general relativity, 226
Astrophysical Journal Letters, The, 195
atomic clock, 42, 49, 56, 58, 148, 172, 175, 177, 182, 255; *see also* hydrogen maser clock
atomic energy, 245

Baade, Walter, 184
Bardeen, James M., 12
baseline: of interferometer, 81
Bell, Jocelyn, 182, 183
Bender, Peter L., 145
Benedetti, Giambattista, 28
Bergmann, Peter G., 11
Berliner Illustrirte, 11
big bang, 12, 15, 148, 152, 163–65, 169, 226, 242; *see also* cosmic fireball radiation; cosmological model; cosmology; universe
binary pulsar, 15, 195, 226, 232, 234; accuracy of measurements in, 198; and gravitational red shift, 199, 205; and gravitational waves, 15, 203–5, 234; and Mercury, compared, 194–95, 196; and relativistic effects, 194–95; and time dilation, 198–99, 205; average separation in, 194; decrease of orbital period of, 204, 205; discovery of, 186–91; distance from Earth, 201; formation of, 200; individual masses in, 200, 205; orbital period of, 191, 198; orbital velocity of, 194; periastron of, 192; periastron shift of, 195–97, 198; pulse period of, 186, 197, 198; total mass of, 197

binary system, 188–89; and gravitational waves, 203–4, 240, 241
binary X-ray system: containing a black hole, 9, 233; containing a neutron star, 234; Cygnus X1, 231–33
black hole, 9, 12, 15, 16, 66, 146, 156, 168, 169, 195, 200, 201; and Cygnus X1, 231–33; and dragging of inertial frames, 230; and gravitational waves, 236, 241, 242; and LMC X3, 233; event horizon of, 229; gravitational field of, 229–30; history of, 66, 226–29; oscillation of, 242; rotating, 229; search for, 230–34; supermassive, 230, 234; *see also* gravitational collapse; Kerr solution; Schwarzschild solution
Bode, Johann, 67
Bolyai, Wolfgang, 22
Bondi, Hermann, 163
Bozeman, John, 135
Bozeman, Montana, 135
Braginsky, Vladimir, 32; and Eötvös experiment, 32–33
Brahe, Tycho, 89
Brans, Carl H., 6, 101, 147, 153, 154, 156; *see also* Brans-Dicke theory
Brans-Dicke theory, 6, 13, 14, 16, 17, 40, 104, 137; and deflection of light, 74, 155, 157; and general relativity, predictions compared, 155; and gravitational redshift, 49, 155; and Mach's principle, 154; and Mercury's perihelion shift, 101, 155, 156, 158; and Nordtvedt effect, 138, 140, 142, 146, 158; and Occam's razor, 159; and solar oblateness, 101, 156, 158; and strong equivalence principle, 138; and time delay of light, 117, 155, 157; and values of ω, 154; and varying gravitational constant, 153–54, 155, 170, 180; decline of, 157–58; impact of, on experimental relativity, 157; limits on ω for, 157–58; reception of, by physicists, 155–57
Bureau International de l'Heure, 255

California Institute of Technology, 4, 12, 16, 154
Caltech, *see* California Institute of Technology
Cambridge University, 75, 182
Cannon, Robert H., 210, 219, 221
celestial mechanics, 91, 93, 95, 126
Center for Astrophysics, 95, 209
center of mass, 188, 191
centrifugal force, 30, 97, 99, 105, 150, 218
Ceres, 179
CERN, 255
cesium-beam clock, 56; *see also* atomic clock; jet-lagged clocks experiment
Chandrasekhar, Subrahmanyan, 10, 12, 200
Chandrasekhar mass, 200, 233
chemical laws: and special relativity, 245, 257
Chinguetti Oasis, Mauritania, 79
Clemence, Gerald M., 175
compact object, 8, 200; in binary X-ray systems, 231–33; *see also* black hole; gravitational collapse; neutron star; white-dwarf star
computer, 14, 95, 123, 132, 242
continental drift, 144
Copernicus, Nicholas, 11, 161
corpuscular theory of light, 66, 67
cosmic fireball radiation, 9, 148, 164; *see also* big bang; cosmology; universe
cosmic rays, 245, 255
cosmological model, 16, 162, 170; static, 162; steady-state theory, 163, 172; *see also* big bang; universe
cosmological term, 162
cosmology, 162–65, 169, 242; *see also* big bang; universe
Crab nebula, 80, 184
Crab pulsar, 184, 186
Crommelin, Andrew, 76
curved space, *see* space, curved; space curvature
curved space-time, *see* space-time, curved
Cygnus X1, 231–33

day, increasing length of, 170–72
de Sitter, Willem, 75, 211
de Sitter effect, 213

Deep-Space Network, 125
deflection of light, 13, 16, 94, 149, 204, 229; and Eddington, 10–11, 75–78; and POINTS, 209; and principle of equivalence, 69–71; and radio interferometers, 81–82; and Soldner, 67, 79; and space curvature, 71–74; and time delay of light, 112, 115, 117, 121; eclipse expeditions, 10, 75–78, 79–80, 86, 87, 88; effect of solar corona on, 85–86; from quasars 82–85; higher-order correction in, 209; in Brans-Dicke theory, 155; in general relativity, 71–74; in Newtonian gravitation theory, 67; over entire sky, 86
deuterium, 169
Dicke, Robert H., 6, 139, 140, 147–49, 217; and Eötvös experiment, 17–18, 32–33; and lunar laser ranging, 143, 145; and Mach's principle, 149–53, 170; and oblateness of Sun, 96–101, 103; and varying gravitational constant, 161, 171–72; *see also* Brans-Dicke theory; Eötvös experiment; oblateness of Sun
Dicke-Goldenberg experiment, *see* oblateness of Sun
dimensionless numbers, 165
Dirac, Paul A. M., 165, 257; and large numbers hypothesis, 165–67, 170
Doppler-cancelation scheme, 61–62
Doppler shift, 7, 46, 47, 51; and orbit of binary pulsar, 191–94, 196, 197, 198, 199; in binary-star systems, 188–89, 232; in Pound-Rebka experiment, 52, 53; in rocket gravitational redshift experiment, 60–62
double quasar, 86
downlink, 60, 133
dragging of inertial frames, 213–16; and absolute rotation, 216–19; and Kerr black hole, 230; and planetary orbits, 220; *see also* precession of gyroscope
Dyson, Frank, 75, 77

Earth: age of, 163, 171; and astronomical unit, 4; and tidal friction, 90; effect on Mercury's perihelion shift, 92; gravitational binding energy of, 137, 141; gravitational radius of, 227; magnetic field of, 185; oblateness of, 97; perihelion shift of, 95–96, 104; speed in orbit, 194, 251
Earth, rotation of, 30, 32, 54, 55, 86, 173–74, 213, 215, 222; and tides, 172–74, 220; and varying gravitational constant, 170, 174
Earth-Moon system: and varying gravitational constant, 171–72; conservation of angular momentum in, 174; precession of, 213; *see also* Earth; gravitational constant, varying; Moon; Nordtvedt effect
eccentric orbit, 192; *see also* elliptical orbit
eccentricity, 192
eclipse expedition, *see* deflection of light
eclipse of Sun, *see* solar eclipse
Eddington, Arthur Stanley, 10–11, 75, 88, 111, 202, 211, 213, 217; and deflection of light, 75–78
Einstein, Albert: and cosmological term, 162–63; and doubling of light deflection, 71; and Eötvös experiment, 31–32; and gravitational waves, 201–2; and Mach's principle, 150; and Mercury's perihelion shift, 93; and Michelson-Morley experiment, 251; and Schwarzschild solution, 227, 228; and Soldner, 67; and three tests of general relativity, 13, 120, 207; and "12 wise men", 10; and unified field theory, 11, 201; and "varying speed" of light, 111; as celebrity, 11, 65–66; as Jew, 78; birth of legend, 11; centenary of birth, 15, 205, 206; confidence of, 88, 107
electromagnetism, 11, 23, 27, 44*n*, 149, 166, 205, 257; and principle of relativity, 247–48
elementary particles, 27, 165, 225, 246, 255, 257
elliptical orbit, 67, 68, 104, 105, 142, 192
Eötvös, Roland von, 18, 29
Eötvös experiment, 17, 29–31, 63, 136, 137, 149, 159; and Brans-Dicke theory, 155; and lunar laser-

Eötvös experiment *(continued)*
ranging experiment, 146; and neutrons, 34; in orbit, 34; Moscow version, 32–33; Princeton version, 18, 32–33, 100
equivalence of mass and energy, 257
equivalence principle, 20, 243; and Brans-Dicke theory, 155; and curved space-time, 20, 24–27, 71, 149; and deflection of light, 67–71, 72; and Eötvös experiment, 29–33; and freely falling laboratory, 25–27, 45, 47, 69, 112, 114; and gravitational energy 136–37; and gravitational red shift, 44–49, 63, 199; and gravitational waves, 235; and internal energy, 33; and neutrons, 34; and Nordtvedt effect, 140, 157; and time delay of light, 113–14, 117; distinguished from general relativity, 38–40, 63; of Einstein, 20; of Newton, 28; pendulum experiments, 29; strong, 138, 146
Euclid, 20, 22
Euclidean space, 20, 24, 25, 26, 116, 117, 118
event horizon, 229

Fairbank, William M., 210, 219, 220, 221
field equations, 40, 162, 227, 228
field of force, 23
FitzGerald, George F., 256
flat space, *see* Euclidean space
flat space-time, *see* space-time, flat
Flatland (Abbott), 24
"Fourth Test of General Relativity," 120
freely falling laboratory, 20, 25, 26, 27, 235, 246; *see also* equivalence principle; gravitational red shift
Freundlich, Erwin, 69
fusion, nuclear, 164, 168, 169

Galilei, Galileo, 20, 28, 138
gamma rays, 52, 53, 57, 183
Gauss, Carl Friedrich, 22
gedanken experiment, 20; and deflection of light, 69–73; and gravitational red shift, 44–49; and simultaneity, 251–53; and time delay of light, 113–17; and time dilation, 253–55
general relativity, 40; alternatives to, 6, 13; and anti-relativity program, 78–79; and black holes, 226–30; and Brans-Dicke theory, predictions compared, 155; and cosmology, 162–65; and curved space-time, 40, 63; and dragging of inertial frames, 213–16; and gravitational constant, 153, 160, 161, 170; and gravitational waves, 202–3, 234–36; and mainstream of physics, 10; and mass of binary pulsar, 197; and Mercury's perihelion shift, 93–94, 107; and neutron stars, 233; and Occam's razor, 158–59; and strong equivalence principle 138, 146; as tool in astronomy, 87–88, 197, 199, 226, 242; as tool in gravitational-wave astronomy, 240–42; deflection of light in, 71–74; difficulty of, 12; Einstein's confidence in, 88, 107; Einstein's three tests of, 13, 120, 207; stagnation of, 11; time delay of light in, 111–19
geodesic, 21, 27, 35
geodetic effect, *see* precession of gyroscope
gigahertz, 132
Gold, Thomas, 163
Goldenberg, H. Mark, 58, 96; *see also* oblateness of Sun
Goldstack, 82
Goldstone telescope, 82, 83, 85
gravitational binding energy, 137, 140
gravitational collapse, 8, 226, 236, 241
gravitational constant, 152, 160, 194, 227; and Brans-Dicke scalar field, 153–54; and general relativity, 153, 160, 161, 170; and Mach's principle 152–53, 161
gravitational constant, varying: and anthropic principle, 167–69; and asteroids, 179; and Brans-Dicke

theory, 153–54, 155, 170, 180; and Dirac, 165; and Earth rotation rate, 170; and large numbers hypothesis, 165–67; and lunar orbit, 171–72, 175–77; and Mach's principle, 6, 152–53, 161, 170; and planetary orbits, 143, 171, 175, 177; results from planetary radar, 178; results from Viking, 178; Van Flandern's measurements, 177
gravitational force, 25, 30, 32, 44, 90, 94, 100, 105, 156, 166, 201, 203, 220, 222, 232, 235
gravitational lens, 87, 226
gravitational mass, 31
gravitational potential, 23, 51
gravitational radius, 227; *see also* black hole
gravitational red shift, 5, 13, 16, 43, 67, 111, 113; and Brans-Dicke theory, 155; and equivalence principle, 44–50; and gravitational collapse, 227; and jet-lagged clocks experiment 54–57; and Pound-Rebka experiment, 5, 52–54, 57; and rocket red shift experiment, 42–43, 60–63; and Sun, 50–51; and time delay of light, 114; and white-dwarf stars, 51–52; effect on clock rates, 48; effect on elapsed time, 50; higher-order correction in, 209; in binary pulsar, 199, 205; universality of, 48; *see also* equivalence principle; space-time, curved
gravitational waves, 16, 201–3; and binary pulsar, 203–5; search for, 234–42; sources of, 235–36, 240–42; Weber's experiments on, 203, 236–39
gravitational-wave astronomy, 234, 242
gravitational-wave detector: improvements on Weber detector, 239–40; laser interferometers, 239–40; of Weber, 203, 236–38
great circle of sphere, 21
"guest star," 184; *see also* supernova
gyroscope, 152, 210; and absolute rotation, 216–19; superconducting, 220; *see also* precession of gyroscope
gyroscope experiment, 209–24; *see also* precession of gyroscope

Hafele, J. C., 56
Hafele-Keating experiment, *see* jet-lagged clocks experiment
Halley, Edmund, 90
Halley's comet, 90
Hartle, James B., 12
Hawking, Stephen W., 12
Haystack telescope, 82, 85, 109, 119, 120, 123, 128, 178
HDE 226868, 232; *see also* Cygnus X1
helium: in stars, 164, 168; liquid, 130, 220, 223; production in early universe, 164, 169
Helmholtz, Hermann, 205
Hertz, Heinrich, 205
Hewish, Antony, 182, 183
Hill, Henry, 102, 103
horizon, *see* event horizon; black hole
Hoyle, Fred, 163, 170, 175
Hoyle-Narlikar theory, 170, 175
Hubble, Edwin, 8, 161, 162
Hulse, Russell A., 181, 182, 195; and discovery of binary pulsar, 186–91
Hulse-Taylor pulsar, *see* binary pulsar
Huntley, Chet, 135
hydrogen, 59, 80, 85, 164, 169
hydrogen maser clock, 58–59, 62, 209

inertia, 28, 149–50, 183; increase in, 246, 256
inertial frame, 25, 54, 246–47, 253; and speed of light, 110, 248, 251; dragging of, 213–19, 220, 230; equivalence of, 217, 247, 256
inertial mass, 31, 94
inferior conjunction, 4, 119, 120, 121
interference fringe, 81, 240, 249
interferometer: as gravitational-wave detector, 239–40; of Michelson, 249, 250; optical, 209; radio, 81–82, 83, 84, 86, 213
inverse square law, 90, 91, 100; modification of, 93, 100, 103, 104
iron-57 (Fe^{57}), 52, 53, 58
Jansky, Karl, 80

Jefferson Tower, Harvard, 52
Jet Propulsion Laboratory, 4, 17, 95, 120, 124, 125, 126, 128, 131, 178, 220
jet-lagged clocks experiment, 54–57
Jewish science, 78
Johnson, Paul, 65
JPL, *see* Jet Propulsion Laboratory
Jupiter, 8, 91, 95, 105, 106, 179

Kant, Immanuel, 172
Keating, Richard, 56
Kepler, Johannes, 11, 89
Kepler's laws, 89, 90, 161
Kerr, Roy P., 228
Kerr solution, 228, 229, 230; *see also* black hole
kilohertz, 238
Kleppner, Daniel, 58

Laplace, Pierce Simon, 66, 226
Large Magellanic Cloud, 233
large numbers hypothesis, 165–67, 170; and anthropic principle, 167–70; and varying gravitational constant, 167
laser, 14, 140, 148, 239
laser interferometer, 239–40
laser ranging, *see* lunar laser ranging
laser retroreflector, *see* retroreflector
Le Verrier, Joseph, 90, 91, 92, 95, 207
Leaning Tower of Pisa, 20, 28, 138
Lenard, Philipp, 78
lens, gravitational, *see* gravitational lens
Lense, J., 220
Levine, Martin, 58, 61
Lick Observatory, 145
limb of Sun, 96
Lincoln Laboratory, 4, 95, 110, 123, 128
little green men, 182
LMC X3, 233
Lobachevski, Nikolai, 22
Lodge, Oliver, 87
London *Times*, 11, 65
Lorentz, Henrik A., 255
Lorentz-FitzGerald contraction, 255
Luna 17 and 21, 145
lunar laser ranging, 140, 143–46, 157, 172
lunar motion, *see* Moon

Mach, Ernst, 149, 150
Mach's principle, 6, 149–50; and Brans-Dicke theory, 154, 156; and Dicke, 152–53; and gravitational constant, 152–53; and Newton's bucket, 150–52; and varying gravitational constant, 6, 152–53, 161, 170
magnetic field: of Earth, 185; of pulsar, 185; of superconducting gyroscope, 223
magnetometer, 223
magnitude of star, 9
Manhattan Project, 226
Mariner program, 14, 133; *see also specific missions*
Mariner 4, 124
Mariner 5, 124
Mariner 6, 124–27, 131, 157
Mariner 7, 124–27, 130, 131, 157
Mariner 8, 124, 130
Mariner 9, 124, 130, 131, 157, 178
Mars, 14, 17, 92, 95, 104, 108, 109, 122, 124, 130, 178, 179; and Mariners 6 and 7, 125–27, 130; and Shapiro time delay, 111, 114, 116, 117, 127–28, 130, 133; and Viking, 130–34
maser, 14, 148, 236; *see also* hydrogen maser clock
mass-energy equivalence, 256–57
Massachusetts Institute of Technology, 4, 95, 110, 131, 136, 146, 148, 210, 221
Matthews, Thomas, 6
Maxwell, James Clerk, 205, 248
Maxwell's equations, 249
Maxwell's theory of electromagnetism, 234; *see also* electromagnetism
McDonald Observatory, 145
megahertz, 59
Mercury, 13, 78, 88, 91, 104, 108, 119, 120, 123, 128, 178, 180; and

binary pulsar, compared, 194–95; and Shapiro time delay, 124, 128, 157; and varying gravitational constant, 178; superior conjunction of, 123, 124; *see also* perihelion shift of Mercury
Merritt Island, 43, 61
metric tensor, 154, 159, 227
Michell, John, 66, 226
Michelson, Albert A., 249
Michelson interferometer, 249
Michelson-Morley experiment, 249–51, 256
microarcsecond, 209
microsecond, 52, 111
microwave, 148, 149
Milky Way galaxy, 162
milliarcsecond, 213, 222
millisecond, 81
Millstone Hill radar antenna, 4, 178
Minkowski, Hermann, 22
Misner, Charles, W., 16
MIT, *see* Massachusetts Institute of Technology
Montana State University, 135, 136
month, increase of, 171, 172, 174; Van Flandern's measurements, 175–77
Monthly Notices of the Royal Astronomical Society, 211–12
Moon, 14, 37, 109, 136, 144, 155, 166, 179, 222, 243; and Apollo 11, 144; and Nordtvedt effect, 140–42, 145–46; and strong equivalence principle, 139; and tidal friction, 172–77; and varying gravitational constant, 171–72, 175–77; librations of, 144; motion of, 90–91; secular acceleration of, 90; *see also* Earth-Moon system; gravitational constant, varying; lunar laser ranging; month, increase of; Nordtvedt effect
Morley, Edward W., 249
Moscow State University, 32–33
Mossbauer, Rudolph, 53
Mossbauer effect, 53
Mt. Palomar, 6
mu meson, *see* muon
Muhleman, Duane O., 120, 121
Mulholland, J. Derral, 145
Mullard radio telescope, 85
muon, 246, 255

nanosecond, 37, 57
Narlikar, Jayant, 170
NASA, *see* National Aeronautics and Space Administration
National Aeronautics and Space Administration, 17, 42, 43, 58, 59, 106, 124, 125, 126, 127, 131, 133, 144, 178, 180, 221, 222, 223
National Radio Astronomy Observatory (NRAO), 82, 85, 184
Nature, 182
nebula, 162, 184, 185
Neptune, 90
neutron, 34, 183; and Eötvös experiment, 34
neutron star, 12, 16, 146, 156, 168, 169, 195, 198, 200, 201, 232, 233, 234, 242; as model for pulsars, 183–86; mass of, 200; maximum mass of, 233; *see also* binary pulsar; pulsar; supernova
New York Times, 10
Newcomb, Simon, 93, 95
Newton, Isaac, 11, 23, 28, 29, 66, 74; and the equivalence principle, 28; *see also* Newton's bucket; Newtonian gravitation theory
Newton's bucket, 150–52, 217–18
Newtonian gravitation theory, 23, 89–90, 111, 122, 123, 160, 161, 191; and absolute rotation, 217; and deflection of light, 66, 67, 71, 74; and first approximation to general relativity, 93–94; and gravitational waves, 201; and oblateness of Sun, 100; and Occam's razor, 158–59; in curved space-time language, 159; orbits in, 67, 68; success of, 89–90
niobium, 223
Nobel Price, 53, 78, 148, 149, 183, 200
Nordtvedt, Kenneth, 136, 146, 157; and Dicke, 139–40; calculations of, 136–38

Nordtvedt effect, 140, 145–46, 158; and lunar orbit, 140–42
Novikov, Igor D., 12
null experiment, 31

oblateness of Sun: and equatorial brightness, 96–97; and planetary orbit perturbations, 104; and solar vibrations, 102–3; and Starprobe, 105–7, 209; consequences for general relativity, 100, 156, 158; Dicke-Goldenberg measurements, 96–99, 147, 149, 156, 158; Dicke's 1983 measurements, 103; effect on perihelion shift, 100–101, 103–4, 156, 243; Hill's measurements, 102–3
observable quantities, 10, 50, 113, 114, 119, 202–3
Occam's razor, 158–59, 180
occulting disk, 96, 97, 98, 99
omega (ω): in Brans-Dicke theory, 154, 155, 156; observational limits on, 157–58
Oppenheimer, J. Robert, 226, 227, 228
opposition, 125
Owens Valley telescope, 82, 83

Penrose, Roger, 5, 9, 12
Pentagon, 221
Penzias, Arno, 148, 164
periastron, 188, 192, 193, 199
periastron shift, 195; and mass of binary pulsar system, 197, 199, 226; of binary pulsar, 195–96, 198
perihelion, 91, 104, 192
perihelion shift: in general relativity, 93–94
perihelion shift of Earth, 95, 104
perihelion shift of Mercury, 13, 78, 88, 92, 93, 95, 105, 108, 120, 149, 179, 195, 196, 207, 209, 243; alternative explanations of, 92–93, 94; and Brans-Dicke theory, 101, 155, 156, 158; and inverse square law, 93, 100; and oblateness of Sun, 100–101, 103, 156; and periastron shift of binary pulsar, compared, 196; effect of other planets, 91–92
periodic table of the elements, 245
Philiponos, Ioannes, 28
Philosophiae Naturalis Principia Mathematica (Newton), 28
photodetector, 96
photon, 44*n*
Physical Review, The, 226
Physical Review Letters, 5, 120, 123
Physics Today, 219
Planck's constant, 165
planetary motion: Kepler's laws of, 89, 90, 161
POINTS, 209
post-Newtonian corrections, 94
postdiction, 169
Pound, Robert V., 5, 53
Pound-Rebka experiment, 5, 14, 52–54, 57
Pound-Rebka-Snider experiment, *see* Pound-Rebka experiment
precession of gyroscope: caused by spurious torques, 220; dragging of inertial frames, 213, 216, 218–19, 220, 224; geodetic effect, 210, 211–13, 216, 220, 224; in Sphereland, 211; Schiff's calculations, 220; Stanford experiment, 221–24
precession of perihelion, *see* perihelion shift
Precision Optical Interferometry in Space (POINTS), 209
Princeton University, 6, 12, 18, 32–33, 79, 102, 148
Principe, Spanish Guinea, 76
principle of equivalence, *see* equivalence principle
principle of relativity, 246, 247–48; and speed of light, 248–49, 251; *see also* special relativity
Ptolemy, 161
Pugh, George E., 221
pulsar, 9, 13, 15, 181, 234; discovery of, 182–83; Hulse-Taylor search for, 186; nature of, 183–86; period, stability of, 183, 186, 197–98; *see also* binary pulsar; neutron star
Pythagorean theorem, 249, 254

quantum chromodynamics, 257
quantum electrodynamics, 257
quantum mechanics, 14, 27, 44*n*, 165, 210, 225, 245, 246, 257
quantum theory, *see* quantum mechanics
quasar, 7–8, 13, 15, 16, 80, 82, 232; and black holes, 228, 229, 230–31; and deflection of light, 82–85, 112, 121, 122; double, 86–88, 226; 0111 + 02, 0116 + 08 and 0119 + 11, 85; 3C48, 6, 7, 8; 3C273 and 3C279, 83–85
quasi-stellar radio source, *see* quasar

radar, 4, 14, 80, 107, 108–9, 128
radar echo, *see* radar ranging
radar ranging, 14, 17, 95, 104, 108–10, 112, 119, 121, 123–24, 126, 128, 133, 145, 157, 178; *see also* time delay
radiation pressure, 129
radio astronomy, 59, 80, 183
radio interferometer, 81–82, 83, 86, 209, 213
radio source, 6, 80, 82, 182, 186
radio telescope, 4, 9, 80, 119, 181, 186
radio waves, 4, 8, 48, 59, 80, 82, 85, 124, 128, 132, 182, 185, 206; deflection of, 80; *see also* binary pulsar; deflection of light; pulsar; time delay
Ramsey, Norman, 58
Rebka, Glen A., 5, 53; *see also* Pound-Rebka experiment
red shift: and expansion of universe, 8; *see also* Doppler shift; gravitational red shift; time dilation
refraction of light, 77, 85, 96
Reichley, Paul, 120, 121
relativistic astrophysics, 9, 12, 15, 16; *see also* Texas Symposium on Relativistic Astrophysics
rest mass, 256
retroreflector, 140, 143, 144, 145; *see also* lunar laser ranging
Ride, Sally K., 19, 20, 34, 35
Riemann, Georg Friedrich, 22
Riemann tensor, 9
Rigel, 223
Rosenthal-Schneider, Ilse, 88
rotation, 30, 108, 150, 166, 213; absolute, 150–52, 217–19; and black holes, 229, 230; and gyroscopes, 210; and pulsars, 183–86; retrograde, of Venus, 108 ; *see also*, dragging of inertial frames; Earth, rotation of; Sun
Royal Greenwich Observatory, 75, 80, 177
Ryle, Martin, 183

S-band, 132, 133
Sandage, Allan, 6
Santa Catalina Laboratory for Experimental Relativity by Astrometry (SCLERA), 100
scalar field, 153, 154, 155, 159
scalar-tensor theory, 6, 147, 153, 154, 156, 217; *see also* Brans-Dicke theory
Schiff, Leonard I., 210, 219, 220, 221, 222
Schild, Alfred, 12
Schwarzschild, Karl, 227
Schwarzschild radius, 227; *see also* black hole
Schwarzschild solution, 227–28, 229; *see also* black hole
Scientific American, 17
scintillation, 182
Scout D rocket, 42, 43, 60, 61, 62, 63; *see also* gravitational red shift
second of arc, *see* arcsecond
secular acceleration, 90; *see also* Moon
seeing, 77
semiconductor, 14
Shapiro, Irwin I., 110, 111, 119, 120, 121, 124, 131, 146, 178
Shapiro time delay, 121, 122, 123, 127, 132, 133, 134, 155, 178, 179; *see also* time delay
simplicity, principle of, *see* Occam's razor
simultaneity, 251
Sirius, companion of, 52
Slipher, Vesto M., 8
Snider, Joseph L., 54

Snyder, Hartland, 226, 227, 228
Sobral, Brazil, 76
solar corona, 85, 106, 132; *see also* deflection of light; time delay
solar eclipse, 10, 68, 76, 82, 96, 176; *see also* deflection of light
solar oblateness, *see* oblateness of Sun
solar system: as arena for testing relativity, 14
solar wind, 99, 129
Soldner, Johann Georg von, 67, 79
space: curved, 20, 22, 25; Euclidean, 20, 25; flat, 20, 25
space curvature: and deflection of light, 71–74; and Mercury's perihelion shift, 94; and time delay, 115–19; in Brans-Dicke theory, 74, 117; rubber-sheet analogy, 117, 118
Space Programs Summary, 120
space shuttle, 19, 34, 38, 39, 107, 223, 234
space-time, 17, 22
space-time, curved, 9, 18, 19, 34, 49, 63, 71, 93, 137, 146, 159; and alternative theories of gravity, 40, 137, 149; and black holes, 227, 229–30; and Brans-Dicke theory, 153, 155; and equivalence principle, 25–27; and everyday effects of gravity, 35–39; and general relativity, 38–40; and gravitation, 24, 27; and gyroscope precession, 210–11
space-time, dragging of, *see* dragging of inertial frames
space-time, flat, 23, 25, 26, 27, 93, 116, 121, 149
space-time continuum, 22
space-time diagram, 35, 36, 39
spacecraft: anchored, 130, 131, 132; and random perturbations, 129–30; *see also* listings under specific spacecraft names
special relativity, 22, 23, 35, 47, 54, 149, 194, 198, 217, 245–57; and freely falling laboratory, 25, 27, 45, 47; and speed of gravitational waves, 201; and speed of light, 37, 66, 248, 251; *see also* inertia; inertial frame; Lorentz-FitzGerald contraction; Michelson-Morley experiment; principle of relativity; simultaneity; speed of light; time dilation
spectral line, 50, 51, 52, 189, 199, 232
spectrum, 6, 7, 8, 86, 230, 232, 257
speed of gravitational waves, 201–2, 235
speed of light, 4, 7, 8, 35, 37, 47, 67, 109, 119, 152, 165, 185, 194, 201, 227, 241, 257; and principle of relativity, 248–51; constancy of, 110, 111, 112, 113, 167, 248–51; *see also* Michelson-Morley experiment; principle of relativity; special relativity
Sphereland, 24, 25, 26, 27, 211, 212
spin: of elementary particles, 257; *see also* rotation
Sputnik, 110, 221
Stanford gyroscope experiment, *see* gyroscope experiment; precession of gyroscope
Stanford University, 19, 35, 136, 210, 219, 220, 221, 222, 224
star: lifetime of, 168; *see also* binary system; white-dwarf star
Stark, Johannes, 78
Starprobe, 105–7, 209
steady-state theory, 163, 164, 165, 170
Stevin, Simon, 28
storage ring, 255
strong equivalence principle, 138, 139, 146
Sun: and equatorial brightness, 97, 102; and gravitational red shift, 50–51; and rapidly rotating core, 99, 103; Arrow to, *see* Starprobe; corona of, *see* solar corona; rotation of, 97; vibrations of, 102–3; *see also* deflection of light; oblateness of Sun; time delay
sunspots, 97
superconducting super collider, 246
superconductivity, 14, 210, 220, 223
superior conjunction, 111, 117, 119, 120, 123, 130, 157; and time delay, 111, 117, 119, 121, 122, 123; of Mariner 6 and 7, 125, 126, 127; of Mars, 130, 133, 134; of Mercury, 123, 128; of Venus, 123, 128
supernova, 168, 184, 200, 236, 240;

and gravitational waves, 241–42; of A.D. 1054, 184, 185; *see also* black hole, gravitational collapse; neutron star; pulsar
Surveyor, 140, 144
synchronization of clocks, 251

Taurus, 184
Taylor, Joseph H., 15, 181, 182, 186, 191, 195, 196, 198, 204, 205; *see also* binary pulsar
tensor, 9, 154, 159, 227
Texas Symposium on Relativistic Astrophysics, 9, 15, 196, 204, 228, 230
Thirring, Hans, 220
Thorne, Kip S., 12, 16, 156
thought experiment, *see gedanken* experiment
tidal force, 25, 222, 232, 235, 236
tidal friction, 90, 172–74, 176; and varying gravitational constant, 176–77
tides: and Earth rotation, 174; and lunar motion, 90, 174; of oceans, 25, 172; of solid Earth, 172
time: and space, in space-time diagram, 37; atomic unit of, 166; *see also* space-time; time delay; time dilation
time delay, 17, 178, 179; and Brans-Dicke theory, 155, 157; and equivalence principle, 113–14; and general relativity, 111, 117, 157; and Mariner, 126–27, 130; and planetary radar, 123–24, 128; and planetary topography, 128–29; and space curvature, 115–19; and spacecraft perturbations, 129; and Viking, 130–34; effect of solar corona, 132–33; method of measurement, 121–23; Muhleman-Reichley calculation, 120–21; Shapiro calculation, 111, 119–20, 121; *see also* Shapiro time delay
time dilation, 246, 253–55; in binary pulsar, 198–99, 205; in jet-lagged clocks experiment, 54–55; in rocket gravitational red shift experiment, 56–63
Times of London, 11, 63
topography of planets, 128–29
torque: in Eötvös experiment, 31; on gyroscope, 220
torsion balance, 29
transponder, 60, 62, 129, 132, 133
21-centimeter line, 59
twin paradox, 54

Uhuru X-ray satellite, 231–32
unification, 225
unified field theory, 11, 201
U.S. Naval Observatory, 56, 95, 124, 175, 177, 255
universe, 7, 150, 154, 168, 242, 249; age of, 163, 164, 166, 167, 169; and Mach's principle, 149–50, 152–53; and Newton's bucket, 150–52; and rotation, 218–19; Einstein's model, 162; evolution of, 12, 15, 155, 164; expansion of, 8, 160, 162, 163, 164, 170, 230; isotropy of, 170; observable, mass of, 152; observable, radius of, 152; steady-state theory of, 163, 164; *see also* big bang; cosmic fireball radiation; cosmology
uplink, 60, 61
Uranus, 90

Van Flandern, Thomas A., 175, 177, 179
Venus, 4, 7, 14, 91, 95, 104, 108, 109, 110, 124, 157, 178; and Shapiro time delay, 119, 120, 124, 128; inferior conjunction of, 4; superior conjunction of, 123, 128
Vessot, Robert F. C., 42, 43, 58, 61, 63
Viking, 14, 17, 130, 131, 133, 157, 178, 179, 180; longevity of, 133–34, 178
Vulcan, 92–93

Wallops Island, 42, 61
wave theory of light, 44*n*, 67
Weapons Sytems Evaluation Group, 221

Weber, Joseph, 203, 236, 240; and reported gravitational-wave detection, 238–39; gravitational-wave detector of, 236–38
Weber bar, 239
weightlessness, 19
Westerbork radio telescope, 85
Wheeler, John A., 12, 16, 229
white-dwarf star, 52, 137, 169, 183, 184, 200, 201, 232, 233; and gravitational red shift, 52; maximum mass of, 200, 233
Wilson, Robert, 148, 164

X-band, 132, 133
X-ray, 48, 183, 200, 231, 233, 234
X-ray astronomy, 19, 231–32
X-ray source, 9, 231, 232, 233

year, increase of: and varying gravitational constant, 171, 175

Zel'dovich, Yaakov B., 12
Zwicky, Fritz, 87, 184